THE FRACTIONAL CALCULUS

This is Volume 111 in
MATHEMATICS IN SCIENCE AND ENGINEERING
A series of monographs and textbooks
Edited by RICHARD BELLMAN, *University of Southern California*

The complete listing of books in this series is available from the Publisher upon request.

THE FRACTIONAL CALCULUS

*Theory and Applications of Differentiation
and Integration to Arbitrary Order*

KEITH B. OLDHAM

Department of Chemistry
Trent University
Peterborough, Ontario

JEROME SPANIER

Department of Mathematics
Claremont Graduate School
Claremont, California

ACADEMIC PRESS   New York and London   1974

A Subsidiary of Harcourt Brace Jovanovich, Publishers

COPYRIGHT © 1974, BY ACADEMIC PRESS, INC.
ALL RIGHTS RESERVED.
NO PART OF THIS PUBLICATION MAY BE REPRODUCED OR
TRANSMITTED IN ANY FORM OR BY ANY MEANS, ELECTRONIC
OR MECHANICAL, INCLUDING PHOTOCOPY, RECORDING, OR ANY
INFORMATION STORAGE AND RETRIEVAL SYSTEM, WITHOUT
PERMISSION IN WRITING FROM THE PUBLISHER.

ACADEMIC PRESS, INC.
111 Fifth Avenue, New York, New York 10003

United Kingdom Edition published by
ACADEMIC PRESS, INC. (LONDON) LTD.
24/28 Oval Road, London NW1

Library of Congress Cataloging in Publication Data

Oldham, Keith B
    The fractional calculus.

    (Mathematics in science and engineering, Vol.)
    Bibliography: p.
    1.   Calculus.   I.   Spanier, Jerome, Date
joint author.   II.   Title.   III.   Series.
QA303.O34   515   73-5304
ISBN 0–12–525550–0

AMS (MOS) 1970 Subject Classifications: 26A33, 44A15,
33A30, 35K05

PRINTED IN THE UNITED STATES OF AMERICA

CONTENTS

Preface   ix
Acknowledgments   xiii

Chapter 1
INTRODUCTION

1.1 Historical Survey   1
1.2 Notation   15
1.3 Properties of the Gamma Function   16

Chapter 2
DIFFERENTIATION AND INTEGRATION TO INTEGER ORDER

2.1 Symbolism   25
2.2 Conventional Definitions   27
2.3 Composition Rule for Mixed Integer Orders   30
2.4 Dependence of Multiple Integrals on Lower Limit   33
2.5 Product Rule for Multiple Integrals   34
2.6 The Chain Rule for Multiple Derivatives   36
2.7 Iterated Integrals   37
2.8 Differentiation and Integration of Series   38
2.9 Differentiation and Integration of Powers   39
2.10 Differentiation and Integration of Hypergeometrics   40

Chapter 3
FRACTIONAL DERIVATIVES AND INTEGRALS: DEFINITIONS AND EQUIVALENCES

3.1 Differintegrable Functions   46
3.2 Fundamental Definitions   47
3.3 Identity of Definitions   51
3.4 Other General Definitions   52
3.5 Other Formulas Applicable to Analytic Functions   57
3.6 Summary of Definitions   59

Chapter 4
DIFFERINTEGRATION OF SIMPLE FUNCTIONS

4.1	The Unit Function	61
4.2	The Zero Function	63
4.3	The Function $x - a$	63
4.4	The Function $[x - a]^p$	65

Chapter 5
GENERAL PROPERTIES

5.1	Linearity	69
5.2	Differintegration Term by Term	69
5.3	Homogeneity	75
5.4	Scale Change	75
5.5	Leibniz's Rule	76
5.6	Chain Rule	80
5.7	Composition Rule	82
5.8	Dependence on Lower Limit	87
5.9	Translation	89
5.10	Behavior Near Lower Limit	90
5.11	Behavior Far from Lower Limit	91

Chapter 6
DIFFERINTEGRATION OF MORE COMPLEX FUNCTIONS

6.1	The Binomial Function $[C - cx]^p$	93
6.2	The Exponential Function $\exp(C - cx)$	94
6.3	The Functions $x^q/[1 - x]$ and $x^p/[1 - x]$ and $[1 - x]^{q-1}$	95
6.4	The Hyperbolic and Trigonometric Functions $\sinh(\sqrt{x})$ and $\sin(\sqrt{x})$	96
6.5	The Bessel Functions	97
6.6	Hypergeometric Functions	99
6.7	Logarithms	102
6.8	The Heaviside and Dirac Functions	105
6.9	The Sawtooth Function	107
6.10	Periodic Functions	108
6.11	Cyclodifferential Functions	110
6.12	The Function $x^{q-1} \exp[-1/x]$	112

Chapter 7
SEMIDERIVATIVES AND SEMIINTEGRALS

7.1	Definitions	115
7.2	General Properties	116
7.3	Constants and Powers	118
7.4	Binomials	120
7.5	Exponential and Related Functions	122
7.6	Trigonometric and Hyperbolic Functions	124
7.7	Bessel and Struve Functions	127
7.8	Generalized Hypergeometric Functions	129
7.9	Miscellaneous Functions	130

Chapter 8
TECHNIQUES IN THE FRACTIONAL CALCULUS

8.1	Laplace Transformation	133
8.2	Numerical Differintegration	136
8.3	Analog Differintegration	148
8.4	Extraordinary Differential Equations	154
8.5	Semidifferential Equations	157
8.6	Series Solutions	159

Chapter 9
REPRESENTATION OF TRANSCENDENTAL FUNCTIONS

9.1	Transcendental Functions as Hypergeometrics	162
9.2	Hypergeometrics with $K > L$	165
9.3	Reduction of Complex Hypergeometrics	166
9.4	Basis Hypergeometrics	168
9.5	Synthesis of $K = L$ Transcendentals	172
9.6	Synthesis of $K = L - 1$ Transcendentals	175
9.7	Synthesis of $K = L - 2$ Transcendentals	177

Chapter 10
APPLICATIONS IN THE CLASSICAL CALCULUS

10.1	Evaluation of Definite Integrals and Infinite Sums	181
10.2	Abel's Integral Equation	183
10.3	Solution of Bessel's Equation	186
10.4	Candidate Solutions for Differential Equations	189
10.5	Function Families	192

Chapter 11
APPLICATIONS TO DIFFUSION PROBLEMS

11.1	Transport in a Semiinfinite Medium	198
11.2	Planar Geometry	201
11.3	Spherical Geometry	204
11.4	Incorporation of Sources and Sinks	207
11.5	Transport in Finite Media	210
11.6	Diffusion on a Curved Surface	216

References 219
Index 225

PREFACE

Students of mathematics early encounter the differential operators d/dx, d^2/dx^2, d^3/dx^3, etc., and some doubtless ponder whether it is necessary for the order of differentiation to be an integer. Why should there not be a $d^{1/2}/dx^{1/2}$ operator, for instance? Or d^{-1}/dx^{-1} or even $d^{\sqrt{2}}/dx^{\sqrt{2}}$? It is to these and related questions that the present work is addressed. It will come as no surprise to one versed in the calculus that the operator d^{-1}/dx^{-1} is nothing but an indefinite integral in disguise, but fractional orders of differentiation are more mysterious because they have no obvious geometric interpretation along the lines of the customary introduction to derivatives and integrals as slopes and areas. The reader who is prepared to dispense with a pictorial representation, however, will soon find that fractional order derivatives and integrals are just as tangible as those of integer order and that a new dimension in mathematics opens to him when the order q of the operator d^q/dx^q becomes an arbitrary parameter. Nor is this a sterile exercise in pure mathematics—many problems in the physical sciences can be expressed and solved succinctly by recourse to the fractional calculus.

Our interest in this subject began in 1968 with the realization that the use of half-order derivatives and integrals leads to a formulation of certain electrochemical problems which is more economical and useful than the classical approach in terms of Fick's laws of diffusion. This discovery stimulated our interest, not only in the applications of the notions of the derivative and integral to arbitrary order, but also in the basic mathematical properties of these fascinating operators. Our collaboration since 1968 has taken us far beyond the original motivation and has produced a wealth of material, some of which we believe to be original. As befits a cooperative effort between a mathematician [J. S.] and a chemist [K. B. O.], our work attempts to expose not only the theory underlying the properties of the generalized operator, but also to illustrate the wide variety of fields to which these ideas may be applied with profit. We do not presume to present an exhaustive survey of the subject, but our aim has been to introduce as many readers as possible to the beauty and utility of this material. Accordingly, we have made a deliberate attempt to keep the mathematical discussions as simple as possible. For example, we have not used techniques of modern functional analysis to deal with d^q/dx^q from an operator-theoretic point of view. This latter approach, which has been taken to some extent by Feller (1952)[1] and Hille (1939, 1948), should prove to be very fruitful but is properly the subject of a much more advanced work. Nor have we sought to incorporate the fractional calculus into the larger field of symbolic, operational mathematics (Boole, 1844; Heaviside, 1893, 1920; Mikusinski, 1959; Friedman, 1969; Bourlet, 1897; Ritt, 1917).

During our investigations of the general theory and applications of differintegrals (a term we have coined to avoid the cumbersome alternate "derivatives or integrals to arbitrary order"), we have discovered that, while this subject is old, dating back at least to Leibniz in its theory and to Heaviside in its application, it has been studied relatively little since the early papers which only hinted at its scope. In the last several years there seems to have taken place a mild revival of interest in the subject, but, in our opinion, the application of these ideas has not yet been fully exposed, primarily because of their unfamiliarity. Our studies have convinced us that differintegral operators may be applied advantageously in many diverse areas. Within mathematics, the subject makes contact with a very large segment of classical analysis and provides a unifying theme for a great many known, and some new, results. Applications outside mathematics include such otherwise unrelated topics as: transmission line theory, chemical analysis of aqueous solutions, design of heat-flux meters, rheology of soils, growth of intergranular grooves at metal surfaces, quantum mechanical calculations, and dissemination of atmospheric pollutants.

In developing the theoretical foundations of the subject our guideline has been to view differintegrals as composing a continuum of operators which include ordinary differentiation and integration, single and multiple, as particular instances. These special cases serve as cornerstones of familiarity which we use to establish credibility for the general properties we study. Thus, after a historical introduction to the subject and a review of facts about the gamma function in Chapter 1, Chapter 2 is devoted to a discussion of the properties of derivatives and integrals to integer order as a reminder of results we shall seek to generalize later. In Chapter 3 we introduce our basic definition (due originally to Grünwald) of a generalized derivative–integral

---

[1] All references cited in this Preface are to works listed in the References, beginning on p. 219. They should not be confused with items in the chronological bibliography appearing at the end of Section 1.1.

and demonstrate the equivalence of this definition with the more familiar one attributed to Riemann and Liouville. It seems time in Chapter 4 to derive and display formulas exemplifying differintegration, and we have chosen simple algebraic functions for this purpose. Chapter 5 deals with the general properties of the differintegral operator which help to clinch the idea that differintegrals of noninteger order are not really so different from ordinary derivatives and integrals.

Chapter 6 is viewed as transitional between the theory of differintegral operators and their application: It deals with the differintegration of certain important functions, including several which find use in later chapters. The applications of differintegration to classical mathematics itself constitute a unit which is presented as Chapters 9 and 10. In the former we demonstrate how differintegration forges powerful links among various transcendental functions, including most of those which arise naturally in mathematical physics. That differintegration can serve as a tool in unifying and extending concepts and techniques encountered in the classical calculus is the message of Chapter 10.

Chapter 11 contains what we regard as the most powerful application of this theory, namely to diffusive transport in a semiinfinite medium. Here a single equation—asserting the proportionality of a second order spatial partial derivative to a first order partial derivative with respect to time—governs a wide variety of transport phenomena: heat in solids, chemical species in homogeneous media, vorticity in fluids, electricity in resistive–capacitative lines, to mention only a few. The replacement of this equation, together with an initial and the asymptotic boundary condition, by an equation linking a first order spatial derivative to a half order temporal derivative is the essence of this application. Inasmuch as it incorporates the initial condition and one of the two boundary conditions, the latter equation represents a halfway house between the problem and its solution. In other words, it describes not one, but an entire class of boundary value problems. Besides the economy engendered by this replacement, the resulting semidifferential equation provides a simple and useful expression for the flux at the boundary, an expression which may be applied even when the boundary condition cannot be expressed as a mathematical function, and which moreover avoids the need for calculations relating to all positions in the medium. The fundamental role played in this theory by semidifferentiation and its inverse explains our preoccupation in Chapter 7 with the results of applying $d^{\frac{1}{2}}/dx^{\frac{1}{2}}$ and $d^{-\frac{1}{2}}/dx^{-\frac{1}{2}}$ to a wide variety of functions. The techniques described in Chapter 8 are likewise valuable tools which we developed primarily to handle the semidifferentiation and semiintegration operators which arise in transport applications.

In summary, then, the book is divided roughly into two parts. The first six chapters deal principally with the general properties of differintegral operators, while Chapters 7–11 are mainly oriented toward the application of these properties to mathematical and other problems.

A word about our intended readership. Our hope is that our book will be readable by, and of interest to, a broad audience. About the only prerequisite is an understanding of the classical calculus, although some familiarity with the ideas involved in solving differential equations would be useful. Among mathematicians our work should be of interest to both classical and functional analysts as well as applications-oriented mathematicians. Because of the diversity of subjects embraced by the fractional calculus, this volume could also interest physical chemists, engineers (electrical, mechanical, and petroleum engineers, among others), and many scientists who have occasion to study transport processes akin to diffusion. We have tried to maintain a readable expository style, being rigorous where it was appropriate, while always striving to tell as complete a story as possible. If we have succeeded, we feel that we have created a book which, while less than a text, is more than a monograph on an old and yet novel subject which makes contact with an amazingly large number of areas of classical and applied mathematics.

It should be stressed that, because fractional derivatives and integrals can always be expressed using ordinary derivatives and integrals (as will be apparent in Chapter 3), any result obtainable through the fractional calculus may also be derived making use only of the concepts and symbolism of classical calculus. Nevertheless, its conceptual elegance and the economy engendered through its use make the fractional calculus much more than a hollow extension of conventional theory.

Writing in 1895 Heaviside said "... the result is a simple fundamental one in fractional differentiation.... But the reader presumably cannot take in the idea of fractional differentiation yet." Our hope in presenting this treatise is to make its readers more willing than Heaviside's to feel at home with the concept of fractional differentiation and integration.

ACKNOWLEDGMENTS

It is a pleasure to record our gratitude to the Science Center of the North American Rockwell Corporation where this work was initiated. During the years of our association there, many colleagues encouraged us, but we particularly thank Drs. E. Richard Cohen, Norman D. Malmuth, and Wayne M. Robertson. Later phases of our work, carried out at Trent University, were generously supported by the National Research Council of Canada. As is usually the case when a book is written, those who suffer the most are the families and the secretaries of the authors. For the arduous task of producing the typescript and diagrams we are indebted to Joan Bailey, Joan Burrett, Pat Cole, Irene Fitzpatrick, Karen Nolte, and Beverly White, and particularly to Kay Asakawa, Penny Dalrymple-Alford, and Alison McLellan. We are pleased, also, to acknowledge the contribution of Professor Bertram Ross to our book. The chronological bibliography on the subject of the fractional calculus which appears at the end of Section 1.1 was prepared by Professor Ross and is printed here with his kind permission.

All scientists necessarily build on the foundations of those who have gone before. We are no exceptions, but our subject is vast and diverse and the distinguished list of our forerunners is correspondingly long. It would be invidious to single out any group of them, nor can we claim to have always done justice to their ideas, but we want our readers to appreciate the magnitude of the contributions of those who have preceded us.

THE FRACTIONAL CALCULUS

CHAPTER 1
INTRODUCTION

1.1 HISTORICAL SURVEY

The concept of differentiation and integration to noninteger order is by no means new. Interest in this subject was evident almost as soon as the ideas of the classical calculus were known—Leibniz (1859) mentions it in a letter to L'Hospital in 1695.[1] The earliest more or less systematic studies seem to have been made in the beginning and middle of the 19th century by Liouville (1832a), Riemann (1953), and Holmgren (1864), although Euler (1730), Lagrange (1772), and others made contributions even earlier.

It was Liouville (1832a) who expanded functions in series of exponentials and defined the qth derivative of such a series by operating term-by-term as though q were a positive integer. Riemann (1953) proposed a different definition that involved a definite integral and was applicable to power series with noninteger exponents. Evidently it was Grünwald and Krug who first unified the results of Liouville and Riemann. Grünwald (1867), disturbed by the restrictions of Liouville's approach, adopted as his starting point the definition of a derivative as the limit of a difference quotient and arrived at definite-integral formulas for the qth derivative. Krug (1890), working through Cauchy's integral formula for ordinary derivatives, showed that Riemann's definite integral had to be interpreted as having a finite lower limit while Liouville's definition, in which no distinguishable lower limit appeared, corresponded to a lower limit $-\infty$.

---

[1] Authors and dates designate entries in our working list of references, to be found at the end of the book. They should not be confused with items in the chronological bibliography appearing at the end of Section 1.1. Cited dates, as in "Leibniz (1859)," refer to the year of publication of the references listed at the end of the book. As with collected works, such a date is not necessarily that of the year of original publication.

Parallel to these theoretical beginnings was a development of the applications of the fractional calculus to various problems. In a sense, the first of these was the discovery by Abel (1823, 1825) in 1823 that the solution of the integral equation for the tautochrone could be accomplished via an integral transform, which, as we shall see, benefits from being written as a semiderivative. A powerful stimulus to the use of fractional calculus to solve problems was provided by the development by Boole (1844) of symbolic methods for solving linear differential equations with constant coefficients. The essence of Boole's idea is the formal expansion of an arbitrary function $f(D)$ of the differential operator as a power series and the solution of differential equations by formal inversion of such series. Boole's methods have subsequently been made rigorous for certain classes of functions f [see Bourlet (1897) and Ritt (1917)] and extended in many directions.

The operational calculus of Heaviside (1892, 1893, 1920), developed by him to solve certain problems of electromagnetic theory, was an important next step in the application of generalized derivatives. Heaviside (1920) introduced fractional differentiation in his investigation of transmission line theory; this concept has been extended by Gemant (1936) for use in problems of elasticity. While Heaviside seemed to scorn the "wet blankets of rigorists," at least some theorists recognized the merit of his techniques, and attempted to justify them by acceptable mathematical standards [see Carson (1926) and Wiener (1926)].

In the present century notable contributions have been made to both the theory and application of the fractional calculus. Weyl (1917), Hardy (1917), Hardy and Littlewood (1925, 1928, 1932), Kober (1940), and Kuttner (1953) examined some rather special, but natural, properties of differintegrals of functions belonging to Lebesgue and Lipschitz classes. Erdélyi (1939, 1940, 1954) and Osler (1970a) have given definitions of differintegrals with respect to arbitrary functions, and Post (1930) used difference quotients to define generalized differentiation for operators $f(D)$, where D denotes differentiation and f is a suitably restricted function. Riesz (1949) has developed a theory of fractional integration for functions of more than one variable. Erdélyi (1964, 1965) has applied the fractional calculus to integral equations and Higgins (1967) has used fractional integral operators to solve differential equations. Other applications include those to rheology (Scott Blair *et al.*, 1947; Shermergor, 1966; Scott Blair, 1947, 1950a,b; Scott Blair and Caffyn, 1949; Graham *et al.*, 1961), to electrochemistry (Belavin *et al.*, 1964; Oldham, 1969a; Oldham and Spanier, 1970; Grenness and Oldham, 1972), to chemical physics (Somorjai and Bishop, 1970), and to general transport problems (Oldham, 1973b; Oldham and Spanier, 1972). The developments are far too numerous to give an exhaustive survey here, nor is this our purpose. The readers interested in further references to the literature may consult the chronological bibliography which ends this section. Virtually no area of classical analysis

1.1 HISTORICAL SURVEY 3

has been left untouched by the fractional calculus. Indeed, could one expect less from the natural extension of perhaps the two most basic operations of mathematics—differentiation and integration?

We close Section 1.1 with an annotated chronological bibliography on fractional calculus prepared by Professor Bertram Ross of the University of New Haven. It is reprinted here intact with his kind permission and is meant to give additional historical perspective to our subject. No attempt has been made to eliminate duplication between Professor Ross' bibliography and our own working references to be found at the end of the book.

Professor Ross has examined many of the papers and texts in the following list and has appended short comments where he deemed it appropriate. His criteria for inclusion were: first investigation of an important development, and frequency of citation.

1695 G. W. Leibniz, Letter from Hanover, Germany, September 30, 1695 to G. A. L'Hospital. *Leibnizen Mathematische Schriften*, Vol. 2, pp. 301–302. Olms Verlag., Hildesheim, Germany, 1962. First published in 1849.

Leibniz wrote prophetically, "Thus it follows that $d^{\frac{1}{2}}x$ will be equal to $x\sqrt[2]{dx:x}$, an apparent paradox, from which one day useful consequences will be drawn."

1697 G. W. Leibniz, Letter from Hanover, Germany, May 28, 1697 to J. Wallis, *Leibnizen Mathematische Schriften*, Vol. 4, p. 25. Olms Verlag., Hildesheim, Germany, 1962. First published in 1859.

In this letter Leibniz discusses Wallis' infinite product for π. Leibniz mentions differential calculus and uses the notation $d^{\frac{1}{2}}y$ to denote a derivative of order $\frac{1}{2}$.

1730 L. Euler, "De Progressionibus Transcentibus, sev Quarum Termini Algebraice Dari Nequeunt." *Comment. Acad. Sci. Imperialis Petropolitanae* **5**, 38–57 (1738).

On p. 55 of "Concerning transcendental progressions whose terms can not be given algebraically," Euler writes, "When n is a positive integer, the ratio $d^n p$, p a function of x, to dx^n can always be expressed algebraically. Now it is asked: what kind of ratio can be made if n be a fraction? If n is a positive integer, d^n can be found by continued differentiation. Such a way, however, is not evident if n is a fraction. But the matter may be expedited with the help of the interpolation of series as explained earlier in this dissertation."

1772 J. L. Lagrange, "Sur une nouvelle espèce de calcul relatif à la différentiation et à l'intégration des quantités variables." *Oeuvres de Lagrange*, Vol. 3, pp. 441–476. Gauthier-Villars, Paris, 1849. First appeared in *Nouv. Mém. Acad. Roy. Sci. Belles-Lett. Berlin* **3**, 185–206 (1772).

Lagrange's contribution in this work is the law of exponents (indices) for operators of integer order:

$$\frac{d^m}{dx^m}\frac{d^n}{dx^n}y = \frac{d^{m+n}}{dx^{m+n}}y.$$

Later, when the theory of fractional calculus started, it became important to know whether this law held true if m and n were fractions.

1812 P. S. Laplace, *Théorie Analytique des Probabilités*, Courcier, Paris, 1820. First appeared in 1812.

On pp. 85 and 186 of the third edition, Laplace writes expressions for certain fractional derivatives.

1819 S. F. Lacroix, *Traité du Calcul Différentiel et du Calcul Intégral*, 2nd ed., Vol. 3 pp. 409–410. Courcier, Paris.

In this 700 page text two pages are devoted to fractional calculus. Lacroix develops a formula for fractional differentiation for the nth derivative of v^m by induction. Then, he formally replaces n with the fraction $\frac{1}{2}$, and together with the fact that $\Gamma(\frac{1}{2}) = \sqrt{\pi}$, he obtains

$$\frac{d^{\frac{1}{2}}}{dv^{\frac{1}{2}}} v = \frac{2\sqrt{v}}{\sqrt{\pi}}.$$

1822 J. B. J. Fourier, "Théorie Analytique de la Chaleur." *Oeuvres de Fourier*, Vol. 1, p. 508. Didot, Paris.

Fourier makes the following generalization:

$$\frac{d^u}{dx^u} f(x) = \frac{1}{2\pi} \int_{-\infty}^{+\infty} f(\alpha)\, d\alpha \int_{-\infty}^{+\infty} p^u \cos\left(px - p\alpha + \frac{u\pi}{2}\right) dp,$$

and states, "The number u will be regarded as any quantity whatever, positive or negative."

1823 N. H. Abel, "Solution de quelques problèmes à l'aide d'intégrales définies." *Oeuvres Complètes*, Vol. 1, pp. 16–18. Grondahl, Christiania, Norway, 1881. This paper first appeared in *Mag. Naturvidenkaberne* (1823).

Abel was probably the first to give an application of fractional calculus. He used derivatives of arbitrary order to solve the tautochrone (isochrone) problem. The integral he worked with

$$\int_0^x (x-t)^{-\frac{1}{2}} f(t)\, dt$$

is precisely of the same form that Riemann used to define fractional operations.

1832 [a] J. Liouville, "Mémoire sur quelques Quéstions de Géometrie et de Mécanique, et sur un nouveau genre de Calcul pour résoudre ces Quéstions." *J. Ecole Polytech.* **13**, Section 21, pp. 1–69.

The first major study of fractional calculus starts with Liouville. On p. 3, Liouville considers $(d^{\frac{1}{2}}/dx^{\frac{1}{2}})e^{2x}$. In this memoir, some problems in mechanics and geometry are solved by the use of fractional operations.

1832 [b] J. Liouville, "Mémoire sur le Calcul des différentielles à indices quelconques." *J. Ecole Polytech.* **13**, Section 21, pp. 71–162.

On p. 94, he considers the existence of a complementary function to be added to the definition of a fractional operation. Liouville was led into error in this particular area. On p. 117, he works out a method for the fractional derivative of a product of two functions.

1832 [c] J. Liouville, "Mémoire sur l'intégration de l'équation $(mx^2 + nx + p)$ $d^2y/dx^2 + (qx+r)\,dy/dx + sy = 0$ à l'aide des différentielles à indices quelconques." *J. Ecole Polytech.* **13**, Section 21, pp. 163–186.

1833 G. Peacock, "Report on the Recent Progress and Present State of Affairs of Certain Branches of Analysis." *Rep. British Assoc. Advancement Sci.* 185–352.

He enunciated the *Principle of the permanence of equivalent forms*, later echoed by Kelland in 1846. Peacock was led into error on the subject of fractional operations by assuming that the above stated principle was valid for all symbolic operations. For example, although $DD^{-1} = D^0$, $D \neq 1/D^{-1}$, where $D = d/dx$.

1834 [a] J. Liouville, "Mémoire sur une formule d'analyse." *J. Reine Angew. Math. (Crelle's Journal)* **12**, 273–287.

Liouville discusses the tautochrone problem.

1834 [b] J. Liouville, "Mémoire sur le théoreme des fonctions complémentaires," *J. Reine Angew. Math. (Crelle's Journal)* **11**, 1–19.

Liouville continues work on the complementary function (1832[b]). He argues that if the differential equation

$$\frac{d^n y}{dx^n} = 0$$

has a complementary solution, why shouldn't

$$\frac{d^u y}{dx^u} = 0$$

have a complementary solution when u is arbitrary?

1835 [a] J. Liouville, "Mémoire sur l'usage que l'on peut faire de la formule de Fourier, dans le calcul des différentielles à indices quelconques." *J. Reine Angew. Math. (Crelle's Journal)* **13**, 219–232.

Liouville suggests a better way of writing Fourier's (1822) formula.

1835 [b] J. Liouville, "Mémoire sur le changement de la variable dans le calcul des différentielles à indices quelconques." *J. Ecole Polytech.* **15**, Section 24, 17–54.

He gives the definition of a fractional derivative, p. 22, as an infinite series:

$$\frac{d^u y}{dx^u} = \sum A_m e^{mx} m^u,$$

where, "u est un nombre quelconque, entier ou fractionnaire, positif ou negativ, réel ou imaginaire."

1839 S. S. Greatheed, "On General Differentiation No. I," *Cambridge Math. J.* **1**, 11–21. In the same issue are two more papers: "On General Differentiation No. II." *Cambridge Math. J.* **1**, 109–117; "On the Expansion of a Function of a Binomial." *Cambridge Math. J.* **1**, 67–74.

In the first two papers above, Greatheed uses Liouville's definition to develop formulas for fractional differentiation. In the third paper, he supplements Taylor's theorem by use of fractional derivatives.

1839 P. Kelland, "On General Differentiation." *Trans. Roy. Soc. Edinburgh* **14**, 567–618.

1841 D. F. Gregory, *Examples of the Processes of the Differential and Integral Calculus*, 1st ed., p. 350, 2nd ed., 1846, p. 354. J. J. Deighton. Cambridge, England.

Gregory was probably the founder of what was then called the *calculus of operations*. He gives the solution of the heat equation

$$\frac{d^2 z}{dx^2} = \frac{1}{a}\frac{dz}{dy}$$

in symbolic operator form:

$$z = A e^{y\beta^{\frac{1}{2}}} + B e^{-y\beta^{\frac{1}{2}}},$$

where $\beta = a^{-1}(d/dx)$. This form was later used by Heaviside.

1842 A. De Morgan, *The Differential and Integral Calculus Combining Differentiation, Integration, Development, Differential Equations, Differences, Summation, Calculus of Variations ... with Applications to Algebra, Plane and Solid Geometry and Mechanics.* Baldwin and Cradock, London, published under the superintendence of the Society for the diffusion of useful knowledge. First published in twenty-five parts.

In this long-titled text, De Morgan devotes three pages to the subject of fractional calculus. He makes the statement that neither the system of fractional operations as given by Peacock nor that of Liouville has any claim to be considered as giving the form $D^u x^n$, though either may be a form. The controversy over different systems of fractional operations was made more explicit by Center, was cleared up in the last decades of the nineteenth century, and was raised again by Post.

1846 P. Kelland, "On General Differentiation." *Trans. Roy. Soc. Edinburgh* **16**, 241–303 (1849).

Kelland assumes that the principle of the *permanence of equivalent forms*, stated for algebra, is valid for all symbolic operations. This principle was used earlier by Peacock, and by G. Boole "On a General Method in Analysis," *Philos. Trans. Roy. Soc. London* **134**, 225–282 (1844), a paper which developed the formal theory of operators. Kelland states, "Algebraic formulae which are the results of these laws and nothing else, must be correct forms also when the algebraic symbols are replaced by such symbols of operation." The mistrust that Heaviside encountered decades later when he submitted his results obtained by the use of symbolic operators might be traced to errors of these mathematicians who misapplied the principle of the *permanance of equivalent forms*.

1847 B. Riemann, "Versuch einer Auffassung der Integration und Differentiation." *Gesammelte Werke*, 1876. ed. publ. posthumously, pp. 331–344; 1892 ed., pp. 353–366. Teubner, Leipzig. Also in *Collected Works* (H. Weber, ed.), pp. 354–360. Dover, New York, 1953.

Riemann sought a generalization of a Taylor's series expansion and derived the following definition for fractional integration:

$$\frac{d^{-r}}{dx^{-r}} u(x) = \frac{1}{\Gamma(r)} \int_c^x (x-k)^{r-1} u(k)\, dk.$$

However, he saw fit to add a complementary function to the above definition. Today, this definition is in common use as a definition for fractional integration but with the complementary function taken to be identically zero, and the lower limit of integration c is usually zero.

1848 C. J. Hargreave, "On the Solution of Linear Differential Equations." *Philos. Trans. Roy. Soc. London* **138**, 31–54.

This paper is notable because it appears to be the first to generalize Leibniz's rule for the nth derivative of a product: $(uv)^{(n)} = (u+v)^{(n)}$, n an integer, generalized to $d^u(uv)/dx^u$, u arbitrary.

1848 [a] W. Center, "On the Value of $(d/dx)^\theta x^0$ When θ Is a Positive Proper Fraction." *Cambridge and Dublin Math. J.* **3**, 163–169.

Using x^0 to denote a constant, unity, Center considers the fractional derivative of x^0. He explicitly defines the controversy over two systems of fractional operations. The system by Peacock

$$\left(\frac{d}{dx}\right)^\theta x^m = \frac{\Gamma(m+1)}{\Gamma(m-\theta+1)} x^{m-\theta}, \qquad \theta > 0,$$

when $m = 0$ yields a finite result. Liouville's system

$$\left(\frac{d}{dx}\right)^\theta x^{-m} = \frac{(-1)^\theta \Gamma(m+\theta)}{\Gamma(m)} x^{-\theta-m}, \qquad \theta > 0, \quad m + \theta > 0,$$

when $m = 0$ equals zero. Center states on p. 166: "The whole question is therefore now plainly reduced to this, what is $(d/dx)^\theta x^0$ when θ is a positive proper fraction? For when this point is settled, we shall have determined at the same time which of the two systems we *must* adopt."

1848 [b] W. Center, "On Differentiation with Fractional Indices, and on General Differentiation." *Cambridge and Dublin Math. J.* 3, 274–285.

1849 W. Center, "On Fractional Differentiation." *Cambridge and Dublin Math. J.* 4, 21–26.

1850 W. Center, "On Fractional Differentiation." *Cambridge and Dublin Math. J.* 5, 206–217.

1855 J. Liouville, "Sur une formule pour les différentielles à indices quelconques à l'occasion d'un Mémoire de M. Tortolini." *J. Math. Pures Appl.* 20, pages unnumbered (1855).

Liouville adds to his discussion of a series definition for a fractional derivative.

1859 H. R. Greer, "On Fractional Differentiation." *Quart. J. Math.* Oxford Ser. 3, 327–330 (1858–1860).

Greer develops formulas for the semiderivatives of $\sin x$ and $\cos x$ using for his starting point Liouville's development $D^{\frac{1}{2}} e^{mx} = m^{\frac{1}{2}} e^{mx}$. He also deals with finite differences of order $\frac{1}{2}$, $\Delta^{\frac{1}{2}}$.

1861 Z. Wastchenxo, "On Fractional Differentiation." *Quart. J. Math.* 4, 237–243.

Additional formulas to those of Greer above are developed.

1865 H. Holmgren, "Om differentialkalkylen med indices af havd natur som helst." *Kongliga Svenska Ventenkaps-Akademiens Handlingar*, Vol. 5, No. 11, 1–83 (1866). (Usually cataloged under Svenska.)

Holmgren took the same integral representation arrived at by Riemann (1847) as his starting point for a monograph on fractional differentiation.

1867 H. Holmgren, "Sur l'intégration de l'équation différentielle

$$(a_2 + b_2 x + c_2 x^2) \, d^2y/dx^2 + (a_1 + b_1 x) \, dy/dx + a_0 y = 0,"$$

Kongliga Svenska Ventenkaps-Akadamiens, Vol. 7, No. 9, 58 pages (1867–1868).

1867 A. K. Grünwald, "Ueber „begrenzte" Derivationen und deren Anwendung." *Z. Math. Phys.* 12, 441–480.

One of three applications is inversion, p. 478. If θ is a known function of x, then by fractional operations one can determine the unknown function $f(t)$ in the integral equation

$$\theta = \int_0^x (x - t)^p f(t) \, dt.$$

1868 [a] A. V. Letnikov, "Theory of Differentiation of Fractional Order." *Mat. Sb.* 3, 1–68.

Letnikov proves for arbitrary orders, pp. 56–58, that:

$$[D^q D^p f(x)]_{x_0}^x = [D^{q+p} f(x)]_{x_0}^x.$$

1868 [b] A. V. Letnikov, "Historical Development of the Theory of Differentiation of Fractional Order." *Mat. Sb.* 3, 85–119.

Letnikov discusses the work of Liouville, Peacock and Kelland.

1872 [a] A. V. Letnikov, "An Explanation of Fundamental Notions of the Theory of Differentiation of Fractional Order." *Mat. Sb.* 6, 413–445.

The main theme here is the generalization of Cauchy's integral formula.

1 INTRODUCTION

1872 [b] A. V. Letnikov, "Studies in the Theory of Integrals of the form $\int_a^x (x-u)^{p-1} f(u)\,du$." *Mat. Sb.* **7**, 5–205 (1874). Summary in French, *Bull. Sci. Math. Astron.* **7**, 233–238 (1874).

In Chapter III, Letnikov applies the theory of fractional calculus to the solution of certain differential equations.

1873 J. Liouville, "Mémoire sur l'intégration des équations différentielles à indices fractionnaires." *J. Ecole Polytech.* **13**, Section 25, pp. 58–84.

1880 A. Cayley, "Note on Riemann's Paper." *Math. Ann.* **16**, 81–82.

Referring to Riemann's paper (1847) he says, "The greatest difficulty in Riemann's theory, it appears to me, is the interpretation of a complementary function containing an infinity of arbitrary constants." The question of the existence of a complementary function caused much confusion. Liouville and Peacock were led into error, and Riemann became inextricably entangled in his concept of a complementary function.

1884 H. Laurent, "Sur le calcul des derivées à indices quelconques." *Nouv. Ann. Math.* [3], **3**, 240–252.

Laurent generalizes Cauchy's integral formula. He does work on the generalized product rule of Leibniz but leaves the result in integral form.

1888 P. A. Nekrassov, "General Differentiation." *Mat. Sb.* **14**, 45–168.

Using Liouville's starting point for pth order differentiation $d^p e^{mx}/dx^p = m^p e^{mx}$, Nekrassov, p. 152, finds the derivative of arbitrary order of $(x-a)^q$.

1890 A. Krug, "Theorie der Derivationen." *Akad. Wiss. Wien Denkenschriften, Math. Naturwiss. Kl.* **57**, 151–228.

1892 J. Hadamard, "Essai sur l'étude des fonctions données par leur développment de Taylor." *J. Math. Pures Appl.* [4], **8**, 101–186.

1892 O. Heaviside, *Electrical Papers.* The Macmillan Company, London.

1893 [a] O. Heaviside, "On Operators in Physical Mathematics." *Proc. Roy. Soc. London* **52**, 504–529 (1893); **54**, 105–143 (1894).

1893 [b] O. Heaviside, Electromagnetic theory, Vol. 1. The Electrician printing and publishing company, ltd., London. Reprinted by Benn, London 1922.

1893 G. Oltramare, *Calcul de Généralization.* Hermann, Paris, reprinted with revisions, 1899, cited by Davis (1936, pp. 94–98).

1899 O. Heaviside, Electromagnetic theory, Vol. 2. The Electrician printing and publishing company, ltd., London. Reprinted by Benn, London 1922.

1902 R. E. Moritz, "On the Generalization of the Differentiation Process." *Amer. J. Math.* **24**, 257–302.

He uses many new symbols and terms making this paper extremely difficult to read.

1902 S. Pincherle, "Sulle derivate ad indice qualunque." *Mem. Reale Accad. Inst. Sci. Bologna* [5], **9**, 745–758, cited by Davis (1936).

1912 O. Heaviside, Electromagnetic theory, Vol. 3. The Electrician printing and publishing company, ltd., London. Reprinted by Benn, London 1922.

1917 G. H. Hardy, "On Some Properties of Integrals of Fractional Order." *Messenger Math.* **47**, 145–150.

1917 H. Weyl, "Bemerkungen zum Begriff des Differentialquotienten gebrochener Ordnung." *Vierteljschr. Naturforsch. Gesellsch. Zurich* **62**, 296–302.

1918 E. Schuyler, Problem #360. *Amer. Math. Monthly* **25**, 173.

What interpretation must be given to $d^{\frac{1}{2}}y/dx^{\frac{1}{2}}$ so that $(d^{\frac{1}{2}}/dx^{\frac{1}{2}})(d^{\frac{1}{2}}y/dx^{\frac{1}{2}}) = dy/dx$? This problem was discussed and solved by Post (1919).

1918 L. O'Shaughnessy, Problem #433. *Amer. Math. Monthly* **25**, 172–173.

Solve the equation $d^{\frac{1}{2}}y/dx^{\frac{1}{2}} = y/x$. This problem was discussed and solved by Post (1919).

1919 E. Post, "Discussion of Problems #360 and #433." *Amer. Math. Monthly* **26**, 37–39.

When two different solutions are presented to Problem #433, Post takes the opportunity to answer Problem #360 at the same time. He explains that the two solutions are correct; however, each solution is based upon a different definition. The proposer, in his solution, used Liouville's definition of integration of fractional order which is equivalent to the definite integral

$$_cD_x^{-v}f(x) = \frac{1}{\Gamma(v)} \int_c^x (x-t)^{v-1} f(t)\, dt$$

with lower limit of integration c being negative infinity, while Post, in his solution, used Riemann's definition, which is the above integral with c equal to zero. Although Post makes no reference to Center (1848[a]), it is clear why Center, with $f(x)$ equal to a constant, would have two different results for the arbitrary derivative.

1919 M. T. Naraniengar, "Fractional Differentiation." *J. Indian Math. Soc.* **11**, 88–95.

In the view of the present writer, Naraniengar makes an unwarranted assumption that enables him to develop the coefficient $R(n)$, namely, $\Gamma(n+1)/\Gamma(n+\frac{1}{2})$, which satisfies the relation

$$D^{\frac{1}{2}}(x^n) = R(n)x^{n-\frac{1}{2}}.$$

1919 T. J. I'a Bromwich, "Examples of Operational Methods in Mathematical Physics." *Philos. Mag.* [6], **37**, 407–419.

He states that the purpose of this paper is to encourage the use of operational methods in the solution of physical problems. In the course of attacking heat and induction problems, Bromwich is led to some general rules which confirm the accuracy of Heaviside's methods, held in doubt for twenty years.

1921 T. J. I'a Bromwich, "Symbolical Methods in the Theory of Conduction of Heat." *Proc. Cambridge Philos. Soc.* **20**, 411–427.

1922 G. H. Hardy, "Notes on Some Points in the Integral Calculus." *Messenger Math.* **51**, 186–192.

Hardy investigates the properties of integrals of fractional order, in particular, theorems of continuity and summability, seeking analogies to properties valid for integer order.

1922 W. C. Brenke, "An Application of Abel's Integral Equation." *Amer. Math. Monthly* **29**, 58–60.

The problem is to determine the shape of a weir notch when the quantity of water through the weir in a given time is a function of the height of the notch. The equation formulated from physical considerations is

$$Q(h) = c \int_0^h (h-t)^{\frac{1}{2}} f(t)\, dt.$$

To determine the unknown function $f(t)$, the process of inversion is simplified by means of fractional operations.

1923 P. Levy, "Sur le dérivation et l'intégration géneralisées." *Bull. Sci. Math.* [2], **47**, Pt. 1, 307–320; Pt. 2, 343–352.

On pp. 317–318, Levy considers the fractional derivative of e^{iz}.

1924 E. J. Berg, "Heaviside's Operators in Engineering and Physics." *J. Franklin Inst.* **198**, 647–702, cited by Davis (1936).

1924 H. T. Davis, "Fractional Operations as Applied to a Class of Volterra Integral Equations." *Amer. J. Math.* **46**, 95–109.

The lack of detailed explanation, understandable in a journal article, is made up for by a review of the theory of fractional calculus before the theory is applied to the solution of certain integral equations. This paper and Davis' 1927 article are, in the view of the present writer, distinguished not only for their contributions to the theory and applications of fractional calculus, but also as examples of how mathematics papers should be written.

1925 E. Stephens, "Bibliography on General (or Fractional) Differentiation." *Washington Univ. (St. Louis) Studies Sci. Ser.* [6], **12**, 149–152.

This bibliography, with some errors in dates and page numbers, has 32 entries without commentary.

1925 G. H. Hardy and J. E. Littlewood, "Some Properties of Fractional Integrals." *Proc. London Math. Soc.* [2], **24**, 37–41.

1927 W. O. Pennell, "A General Operational Analysis." *J. Math. and Phys.* **7**, 24–38.

Theorems on operational calculus are presented. In the course of solving a linear operational equation $p^2 y + xy = 1$, $p = d/dx$, Pennell gives a detailed explanation of how p can be expanded in a series. This results in a particular integral solution of the differential equation $y'' + xy = 1$.

1927 H. T. Davis, "The Application of Fractional Operators to Functional Equations." *Amer. J. Math.* **49**, 123–142.

Davis discusses the various notations used to define fractional operations. His suggestion is the notation $_cD_x^{-v}f(x)$ to define $\int_c^x (x-t)^{v-1} f(t)\, dt / \Gamma(v)$. Properties of fractional operators are reviewed and then applied to the solution of certain differential equations with operators having fractional exponents, for example, $_cD_x^{3/2}u + \lambda u = f(x)$.

1928 G. H. Hardy and J. E. Littlewood, "Some Properties of Fractional Integrals, I." *Math. Z.* **27**, 565–606 (1928); "Some Properties of Fractional Integrals, II." *Math. Z.* **34**, 403–439 (1932).

In part I, their purpose is to develop properties of the Riemann–Liouville integral and derivative of arbitrary order of functions of certain standard classes, in particular the "Lebesgue class L^p." Part II is an extension of the first paper to the complex field.

1930 E. L. Post, "Generalized Differentiation." *Trans. Amer. Math. Soc.* **32**, 723–781.

1931 L. M. Blumenthal, "Note on Fractional Operators, and the Theory of Composition." *Amer. J. Math.* **53**, 483–492. Cited by Davis (1936).

1931 H. T. Davis, "Properties of the Operator $z^{-v} \log z$, where $z = d/dx$." *Bull. Amer. Math. Soc.* **37**, 468–479.

1931 Y. Watanabe, "Notes on the Generalized Derivative of Riemann–Liouville and its Application to Leibniz's Formula." *Tôhoku Math. J.* **34**, 8–41.

1933 K. S. Cole, "Electric Conductance of Biological Systems," *Proc. Cold Spring Harbor Symp. Quant. Biol.* pp. 107–116. Cold Spring Harbor, New York.

The problem is to express analytically the strength of stimulus that, when applied to the nerve bundle, will change the potential difference across the membrane of an individual fiber by a threshold amount in a given time. See also Davis (1936, pp. 288–289).

1935 A. Zygmund, *Trigonometric Series*, Vol. II, 1st. ed. Z. subwencji Funduszu kultury narodewej, Warsaw; 2nd ed. Cambridge University Press, Cambridge, 1959, pp. 132–142.

In the section entitled "Fractional Integration," Zygmund considers a definition of fractional integration introduced by Weyl more convenient for trigonometric series.

1.1 HISTORICAL SURVEY 11

1936 H. Poritsky, "Heaviside's Operational Calculus—Its Applications and Foundations." *Amer. Math. Monthly* **43**, 331–344.

In an exceptionally well-written paper, Poritsky states that is it hoped that this paper will serve to popularize this subject in the United States and that it will induce mathematicians to include it in their research and curricula. He explains how a correct interpretation of the operator $p^{\frac{1}{2}}$, $p = d/dx$, can be obtained by expanding in powers of $p^{\frac{1}{2}}$ and *neglecting the integer powers*. This procedure, discovered by Heaviside, was never justified by him. Operational calculus, says Poritsky, effects a connection between linear functional transformations enjoying the "translational or shifting property" and analytic functions, by expressing such operations as analytic functions of p; it is thus related to the modern developments of transformations in Hilbert space.

1936 H. T. Davis, *The Theory of Linear Operators*, Principia Press, Bloomington, Indiana.

This text contains an extensive bibliography of operator theory pp. 571–616. Davis develops fractional calculus pp. 64–75 and gives applications pp. 276–292.

1936 W. Fabian, "Fractional Calculus." *J. Math. and Phys.* **15**, 83–89.

In this paper some properties of the fractional integral at infinity are studied, and Fabian deduces therefrom a method of summability of series and integrals. He extends Riemann's definition which enables him to perform fractional integrations along any simple curve in the complex plane in contrast to Hardy and Littlewood (1928) who integrated along straight lines in the complex plane.

1938 L. C. Young and E. R. Love, "On Fractional Integration by Parts." *Proc. London Math. Soc.* [2] **44**, 1–28.

1939 A. Erdélyi, "Transformation of Hypergeometric Integrals by Means of Fractional Integration by Parts." *Quart. J. Math. Oxford Ser.* **10**, 176–189.

1940 H. Kober, "On Fractional Integrals and Derivatives." *Quart. J. Math. Oxford Ser.* **11**, 193–211.

In the first part of the paper, Kober extends some results of Hardy and Littlewood (1925) over a wider range. In the second part of the paper he deals with Mellin transforms, and also a uniqueness theorem for a solution to the equation

$$g(x) = \int_a^x (x-t)^{\alpha-1} f(t)\, dt.$$

1941 D. V. Widder, *The Laplace Transform*, pp. 70–75. Princeton Univ. Press, Princeton, New Jersey.

Widder discusses the connection of the Laplace transform with fractional integrals.

1945 A. Zygmund, "Theorem on Fractional Derivatives." *Duke Math. J.* **12**, 455–464.

1949 M. Riesz, "L'intégrale de Riemann-Liouville et le Probléme de Cauchy." *Acta Math.* **81**, 1–223.

This work, in collaboration with Hagstrom, was started in 1933. The fundamental aspects of fractional calculus are given on pp. 10–16. The remainder of the text deals with various aspects of the fractional integral

$$I^v f(x) = \frac{1}{\Gamma(v)} \int_a^x (x-t)^{v-1} f(t)\, dt$$

in the theory of potentials, Lorentz space, relativistic theory, and wave equation in Riemann space.

1 INTRODUCTION

1950 N. Stuloff, "Die Differentiation beliebiger reelen Ordnung." *Math. Ann.* **122**, 400–410.

Differences of fractional order are discussed:

$$\Delta^{\alpha} x_n = \sum_{v=0}^{\infty} (-1)^v \binom{\alpha}{v} x_{n+v}.$$

1953 B. Kuttner, "Some Theorems on Fractional Derivatives." *Proc. London Math. Soc.* [3] **3**, 480–497.

Kuttner considers the relation between the integrals:

$$\frac{d^n}{dx^n} \frac{1}{\Gamma(n-k)} \int_0^x (x-t)^{n-k-1} f(t)\, dt$$

and

$$(-1)^n \frac{d^n}{dx^n} \frac{1}{\Gamma(n-k)} \int_x^1 (t-x)^{n-k-1} f(t)\, dt.$$

1953 I. I. Hirschmann, "Fractional Integration." *Amer. J. Math.* **75**, 531–546.

1954 A. Erdélyi and staff of the Bateman Manuscript Project, *Tables of Integral Transforms*, Vol. 2, pp. 181–214. McGraw-Hill, New York.

A bibliography of twenty entries on p. 184, without commentary, deals mainly with fractional integrals.

1959 J. L. Lions, "Sur l'existence de solutions des équations de Navier–Stokes," *C. R. Acad. Sci.* **248**, 2847–2849.

A weak solution to the Navier–Stokes equations is a function $u(t)$ from the negative real numbers to L^2, saitsfying a certain functional equation. Lions was the first to pose the question as to whether a weak solution possesses a fractional derivative with respect to t. Then he proceeds to show that if the number of dimensions does not exceed four, then, corresponding to any initial data, there is a weak solution with a fractional derivative of any order less than $\frac{1}{4}$. See also Shinbrot (1971).

1960 A. Erdélyi and I. N. Sneddon, "Fractional Integration and Dual Integral Equations." *Canad. J. Math.* **14**, 685–693.

1961 M. A. Bassam, "Some Properties of the Holmgren–Riesz Transform." *Ann. Scoula Norm. Sup. Pisa* [3] **15**, 1–24.

Bassam shows the equivalence between two definite integrals of arbitrary order given by Holmgren and Riesz, and thus establishes one combined definition. Bassam also wrote a dissertation, "Holmgren–Riesz Transforms." Ph. D. Thesis, Univ. of Texas, Austin, June 1951.

1961 R. Courant, *Differential and Integral Calculus* (translated by E. J. McShane), Vol. 2, pp. 339–341. Wiley (Interscience), New York.

A brief exposition of integrals and derivatives of arbitrary order is given. It is a curious fact worthy of mention that these generalized operators have secured only passing references in standard works in calculus.

1961 A. S. Peters, "Certain Dual Integral Equations and Sonine's Integrals." *Tech. Rep.* No. 225, IMM-NYU. Courant Inst. Math. Sci. New York University. Cited by Buschman (1964).

1962 A. Erdélyi, *Operational Calculus and Generalized Functions*, pp. 2–4. Holt, New York.

A brief discussion of the Heaviside operator $(d/dt)^{\frac{1}{2}}$ is given.

1964 A. Erdélyi, "An Integral Equation Involving Legendre Functions." *SIAM J. Appl. Math.* **12**, 15–30.

1.1 HISTORICAL SURVEY 13

1964 T. P. G. Liverman, *Generalized Functions and Direct Operational Methods*, Vol. I, pp. 28–32. Prentice-Hall, Englewood Cliffs, New Jersey.

1964 I. M. Gel'fand and G. E. Shilov, *Generalized Functions*, Vol. 1, pp. 115–122. Academic Press, New York.

Many special functions can be written as derivatives of arbitrary order of elementary functions. Gel'fand and Shilov give two examples, the hypergeometric function and the Bessel function.

1964 R. G. Buschman, "Fractional Integration," *Math. Japon.* **9**, 99–106.

In the analysis of mixed boundary value problems, dual integral equations are often encountered, as for example $\int_0^\infty y^p J_u(xy) f(y)\, dy = g(x)$ and $\int_0^\infty y^s J_v(xy) f(y)\, dy = h(x)$, where J is the usual Bessel function, $g(x)$ and $h(x)$ are given, and $f(x)$ is to be determined. By showing the connection between fractional integral operators and the algebra of functions that have the Mellin convolution as product, Buschman shows that certain identities can be obtained which have previously appeared as special cases. By using fractional operators, he shows how the dual equations above can be reduced to a single integral equation of the type $\int_0^\infty y^t J_\lambda(xy) f(y)\, dy = F(x)$.

1964 V. A. Belavin, R. Sh. Nigmatullin, A. I. Miroshnikov, and N. K. Lutskaya, "Fractional differentiation of oscillographic polarograms by means of an electrochemical two-terminal network." *Tr. Kazan. Aviacion. Inst.* **5**, 144–145.

1964 T. P. Higgins, "A Hypergeometric Function Transform." *SIAM J. Appl. Math.* **12**, 601–612.

1965 T. P. Higgins, "The Rodrigues Operator Transform, Preliminary Report." Document DI-82-0492, Boeing Sci. Res. Lab., Seattle, Washington.

1965 T. P. Higgins, "The Rodrigues Operator Transform, Table of Generalized Rodrigues Formulas." Document DI-42-0493, Boeing Sci. Res. Lab., Seattle, Washington.

1965 L. von Wolfersdorf, "Über eine Beziehung zwischen Integralen nichtganzer Ordnung." *Math. Z.* **90**, 24–28.

1965 A. Erdélyi, "Axially Symmetric Potentials and Fractional Integration." *SIAM J. Appl. Math.* **13**, 216–228.

1966 I. N. Sneddon, *Mixed Boundary Value Problems in Potential Theory*, pp. 46–52. Wiley, New York.

1967 R. N. Kesarwani, "Fractional Integration and Certain Dual Integral Equations." *Math. Z.* **98**, 83–88.

Kesarwani extends the earlier work of Buschman (1964).

1967 T. P. Higgins, "The Use of Fractional Integral Operators for Solving Nonhomogeneous Differential Equations." Document DI-82-0677, Boeing Sci. Res. Lab., Seattle, Washington.

Although results using fractional integral operators, states Higgins, can always be obtained by other methods, the succinct simplicity of the formulation may often suggest approaches not evident in a classical approach. In this paper, some applications to nonhomogeneous differential equations are given.

1967 G. K. Kalisch, "On Fractional Integrals of Pure Imaginary Order in L_p." *Proc. Amer. Math. Soc.* **18**, 136–139.

1968 S. G. Samko, "A Generalized Abel Equation and Fractional Integral Operators." *Differencial'nye Uravnenija* **4**, 298–314.

1968 M. C. Gaer, *Fractional Derivatives and Entire Functions*. Ph.D. Thesis, Univ. of Illinois, Urbana, Illinois.

1968 G. V. Welland, "Fractional Differentiation of Functions." *Proc. Amer. Math. Soc.* **19**, 135–141.

This paper, a portion of Welland's doctoral dissertation at Purdue University, gives some results of a special nature for functions which have lacunary Fourier series. Using Liouville's definition for fractional integration $[1/\Gamma(\beta)] \int_{-\infty}^{x} (x-t)^{\beta-1} f(t) \, dt$, this paper extends some of the work of Zygmund (1935).

1970 M. C. Gaer and L. A. Rubel, "The Fractional Derivative via Entire Functions." *J. Math. Anal. Appl.* **34**, 289–301.

1970 H. Kober, "New Properties of the Weyl Extended Integral." *Proc. London Math. Soc.* [3] **21**, 557–575.

1970 D. M. Bishop and R. L. Somorjai, "Integral-Transformation Trial Functions of the Fractional-Integral Class." *Phys. Rev. A* [3] **1**, 1013–1018.

1970 K. B. Oldham and J. Spanier, "The Replacement of Fick's Laws by a Formulation Involving Semidifferentiation." *J. Electroanal. Chem.* **26**, 331–341.

The operation of order $\tfrac{1}{2}$, $d^{\frac{1}{2}} f(t)/dt^{\frac{1}{2}}$, which these authors have called semi-differentiation, enables the concentration of an electroactive species at the surface of the electrode to be related straightforwardly to the faradaic current density. Through fractional operations, Oldham and Spanier claim to have uncovered a novel method of elucidating electrochemical kinetics.

1970 [a] T. J. Osler, "Leibniz Rule for Fractional Derivatives Generalized and an Application to Infinite Series." *SIAM J. Appl. Math.* **16**, 658–674.

Certain generalizations of the Leibniz rule for the derivative of the product of two functions are examined and used to generate several infinite series expansions relating special functions.

1970 [b] T. J. Osler, "The Fractional Derivative of a Composite Function." *SIAM J. Math. Anal.* **1**, 288–293.

Osler derives a generalized chain rule, and examines a few special cases of this rule.

1971 [a] T. J. Osler, "Taylor's Series Generalized for Fractional Derivatives and Applications." *SIAM J. Math. Anal.* **2**, 37–47.

1971 [b] T. J. Osler, "Fractional Derivatives and Leibniz Rule." *Amer. Math. Monthly* **78**, 645–649.

The fractional derivative is defined by generalizing Cauchy's integral formula. This definition is used to generalize the Leibniz rule for the derivative of a product by means of which the value of the hypergeometric function of unit argument is evaluated in terms of the gamma function.

1971 E. R. Love, "Fractional Derivatives of Imaginary Order." *J. London Math. Soc.* [2] **3**, 241–259.

In the usual definitions of fractional differentiation, the real part of the order of differentiation is restricted to values greater than zero. Love defines fractional differentiation when the order is purely imaginary, $\text{Re}(\alpha) = 0$, in such a way that the properties of the usual definition are extended to the case of pure imaginary order.

1971 M. Shinbrot, "Fractional Derivatives of Solutions of Navier–Stokes Equations." *Arch. Rational Mech. Anal.* **40**, 139–154.

Shinbrot answers the question posed by Lions (1959) and proves that the order of fractional differentiation can be extended to $\tfrac{1}{2}$.

1971 P. Butzer and R. Nessel, *Fourier Analysis with Approximation*, pp. 400–403. Academic Press, New York.

1972 K. B. Oldham, "A Signal-Independent Electroanalytical Method." *Anal. Chem.* **44**, 196–198.

1972 M. Grenness and K. B. Oldham, "Semiintegral Electroanalysis: Theory and Verification." *Anal. Chem.* **44**, 1121–1129.

1972 K. B. Oldham and J. Spanier, "A General Solution of the Diffusion Equation for Semiinfinite Geometries." *J. Math. Anal. Appl.* **39**, 655–669.

1972 [a] T. J. Osler, "A Further Extension of the Leibniz Rule to Fractional Derivatives and Its Relation to Parseval's Formula." *SIAM J. Math. Anal.* **3**, 1–15.

1972 [b] T. J. Osler, "An Integral Analogue of Taylor's Series and Its Use in Computing Fourier Transforms." *Math. Comp.* **26**, 449–460.

1972 [c] T. J. Osler, "Integral Analog of the Leibniz Rule," *Math. Comp.* **26**, 903–915.

1972 R. K. Juberg, "Finite Hilbert Transforms in L_p." *Bull. Amer. Math. Soc.* **78**, 3, 435–438.

1972 T. R. Prabhakar, "Hypergeometric Integral Equations of a General Kind and Fractional Integration." *SIAM J. Math. Anal.* **3**, 422–425.

Some integral equations containing hypergeometric functions in two variables are studied with the use of fractional integration.

1974 To appear in late 1974 or early 1975 B. Ross, "*A Profile of Fractional Calculus*," Chapter V. Ph.D. Thesis, New York Univ., New York.

1975 The Proceedings of the University of New Haven Colloquium on Fractional Calculus and Its Applications to the Mathematical Sciences to be held June 1974 is expected to be published late in 1975.

1.2 NOTATION

Wherever possible we have followed the nomenclature and symbolism of Abramowitz and Stegun's excellent reference text (1964). To avoid unusual fonts, however, we adopt the following symbols:

$S_n^{[m]}$ for Stirling numbers of the second kind (Abramowitz and Stegun, 1964, p. 824).

$H_\nu(\ )$ and $L_\nu(\ )$ for Struve functions and modified Struve functions of order ν (Abramowitz and Stegun, Chapter 12).

We trust that our use of $H_\nu(\ )$ and $L_\nu(\ )$ will not cause confusion with Abramowitz and Stegun's choice of these same symbols for Hermite and Laguerre polynomials (for which we find no use in the present work). The symbolism

$$\left[x \frac{b_1, b_2, \ldots, b_K}{c_1, c_2, \ldots, c_L} \right]$$

and the terminology "generalized hypergeometric function of complexity $\frac{K}{L}$" are introduced and explained in Section 2.10. Its relationship to the usual symbol for the generalized hypergeometric function is also discussed there.

Equation (x.y.z) denotes the zth numbered equation of section y of chapter x. Figures and tables are treated similarly, with the numbering system Figure x.y.z, Table x.y.z used consecutively in each section. Authors' names with publication dates are used to make reference to our bibliography (to be found at the end of the book).

The term "differintegral," by which we mean derivative or integral to arbitrary order, was mentioned in our Preface. We freely make use of the blood relatives of this term: "differintegration," "differintegrand," "semi-integral," etc. Our special symbolism for such operators is introduced and fully explained in the preamble to Chapter 3. Symbols such as f, g denote general differintegrable functions while the symbols ϕ, ψ are reserved for analytic functions.

Few other nonstandard notations are employed. Those that are will be fully defined at their introduction.

1.3 PROPERTIES OF THE GAMMA FUNCTION

The complete gamma function $\Gamma(x)$ plays an important role in the theory of differintegration. Accordingly, it is convenient to collect here certain formulas relating to this function. A comprehensive definition of $\Gamma(x)$ is that provided by the Euler limit

$$\Gamma(x) \equiv \lim_{N \to \infty} \left[\frac{N!\, N^x}{x[x+1][x+2]\cdots[x+N]} \right],$$

but the integral transform definition

(1.3.1) $$\Gamma(x) \equiv \int_0^\infty y^{x-1} \exp(-y)\, dy, \quad x > 0,$$

is often more useful, although it is restricted to positive x values.

An integration by parts applied to the definition (1.3.1) leads to the recurrence relationship

(1.3.2) $$\Gamma(x+1) = x\Gamma(x),$$

which is the most important property of the gamma function. The same result is a simple consequence of the Euler limit definition. Since

$$\Gamma(1) = 1,$$

this recurrence shows that for a positive integer n

(1.3.3)
$$\Gamma(n+1) = n\Gamma(n) = n[n-1]\Gamma(n-1) = \cdots = n[n-1]\cdots 2 \cdot 1 \cdot \Gamma(1) = n!.$$

Rewritten as

$$\Gamma(x-1) = \Gamma(x)/[x-1],$$

1.3 PROPERTIES OF THE GAMMA FUNCTION

the recurrence formula also serves as an analytic continuation, extending the definition of the gamma function to the negative arguments to which definition (1.3.1) is inapplicable. This extension shows $\Gamma(0)$ to be infinite, as is $\Gamma(-1)$ and the value of the gamma function at all negative integers. Ratios of gamma functions of negative integers are, however, finite; thus if N and n are positive integers,

$$(1.3.4) \quad \frac{\Gamma(-n)}{\Gamma(-N)} = [-N][-N+1]\cdots[-n-2][-n-1] = [-]^{N-n}\frac{N!}{n!}.$$

The reciprocal $1/\Gamma(x)$ of the gamma function is single-valued and finite for all x. Figure 1.3.1 shows a graph of this function. Note the continuous alternation of sign for negative argument and the asymptotic approach to zero for large positive x, an approach described by

$$(1.3.5) \quad \frac{1}{\Gamma(x)} \sim \frac{x^{\frac{1}{2}-x}}{\sqrt{2\pi}} \exp(x), \qquad x \to \infty.$$

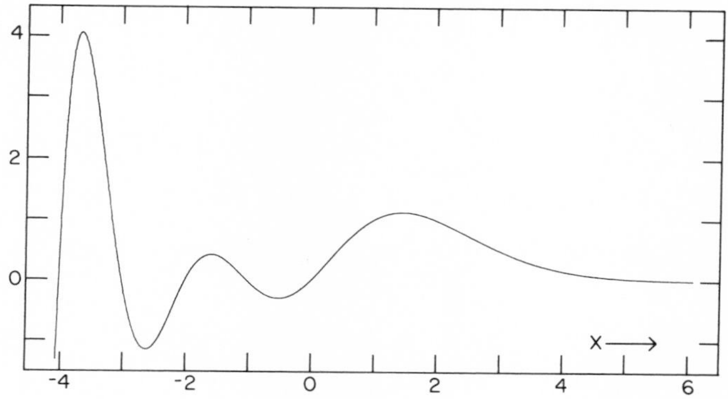

FIG. 1.3.1 The reciprocal, $1/\Gamma(x)$, of the gamma function for $-4 \leq x \leq 6$.

As we have seen, the gamma function of a positive integer n is itself a positive integer, while the gamma function $\Gamma(-n)$ of a negative integer is invariably infinite. The gamma functions $\Gamma(\frac{1}{2}+n)$ and $\Gamma(\frac{1}{2}-n)$ turn out to be multiples of $\sqrt{\pi}$; thus,

$$\Gamma(\tfrac{1}{2}) = \sqrt{\pi},$$

$$(1.3.6) \quad \Gamma(\tfrac{1}{2}+n) = \frac{(2n)!\sqrt{\pi}}{4^n n!},$$

and

(1.3.7) $$\Gamma(\tfrac{1}{2} - n) = \frac{[-4]^n n! \sqrt{\pi}}{(2n)!}.$$

Some frequently encountered examples are included in Table 1.3.1.

Table 1.3.1 *Some values of the gamma function $\Gamma(x)$ for integer and half-integer x*

$\Gamma(-\tfrac{3}{2}) = \tfrac{4}{3}\sqrt{\pi}$	$\Gamma(1) = 1$
$\Gamma(-1) = \pm\infty$	$\Gamma(\tfrac{3}{2}) = \tfrac{1}{2}\sqrt{\pi}$
$\Gamma(-\tfrac{1}{2}) = -2\sqrt{\pi}$	$\Gamma(2) = 1$
$\Gamma(0) = \pm\infty$	$\Gamma(\tfrac{5}{2}) = \tfrac{3}{4}\sqrt{\pi}$
$\Gamma(\tfrac{1}{2}) = \sqrt{\pi}$	$\Gamma(3) = 2$

Two useful properties of the gamma function are its reflection

(1.3.8) $$\Gamma(-x) = \frac{-\pi \csc(\pi x)}{\Gamma(x+1)}$$

and its duplication

(1.3.9) $$\Gamma(2x) = \frac{4^x \Gamma(x) \Gamma(x + \tfrac{1}{2})}{2\sqrt{\pi}},$$

the latter being an instance of the Gauss multiplication formula

(1.3.10) $$\Gamma(nx) = \sqrt{\frac{2\pi}{n}} \left[\frac{n^x}{\sqrt{2\pi}}\right]^n \prod_{k=0}^{n-1} \Gamma\left(x + \frac{k}{n}\right).$$

Table 1.3.2 *Some examples of Stirling numbers of the first kind, $S_j^{(m)}$*

	\multicolumn{6}{c}{m}					
j	0	1	2	3	4	5
0	1	0	0	0	0	0
1	0	1	0	0	0	0
2	0	-1	1	0	0	0
3	0	2	-3	1	0	0
4	0	-6	11	-6	1	0
5	0	24	-50	35	-10	1

1.3 PROPERTIES OF THE GAMMA FUNCTION   19

In Chapter 3 the gamma function expression

(1.3.11)
$$\frac{\Gamma(j-q)}{\Gamma(-q)\Gamma(j+1)}$$

will be encountered, where j is a nonnegative integer and q may take any value. For small numerical values of j, expression (1.3.11) is readily simplified to a polynomial in q by application of rules (1.3.2) and (1.3.3). This procedure generalizes to give

(1.3.12)
$$\frac{\Gamma(j-q)}{\Gamma(-q)\Gamma(j+1)} = \frac{[-]^j}{j!} \sum_{m=0}^{j} S_j^{(m)} q^m,$$

where $S_j^{(m)}$ is a Stirling number of the first kind, examples of which are to be found in Table 1.3.2. These numbers are defined by the recurrence

$$S_{j+1}^{(m)} \equiv S_j^{(m-1)} - jS_j^{(m)}; \qquad S_0^{(m)} = S_j^{(0)} = 0, \qquad \text{except} \quad S_0^{(0)} = 1.$$

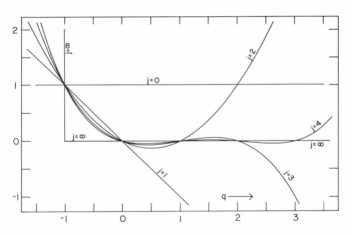

FIG. 1.3.2 The polynomial $\Gamma(j-q)/[\Gamma(-q)\Gamma(j+1)]$ for $j = 0, 1, 2, 3, 4,$ and ∞.

Its expressibility in (1.3.12) as a polynomial in q establishes that expression (1.3.11) is finite and single-valued for all finite values of q and j. Figure 1.3.2 displays the numerical value of the polynomial for real values of q in the range $-2 \leq q \leq 5$ and for $j = 0, 1, 2, 3, 4,$ and ∞.

Equation (1.3.12) provides an expression for the gamma function quotient $\Gamma(j-q)/\Gamma(-q)$. The other quotient $\Gamma(j-q)/\Gamma(j+1)$ appearing in expression (1.3.11) is also of interest. This quotient has the asymptotic expansion

(1.3.13)
$$\frac{\Gamma(j-q)}{\Gamma(j+1)} \sim j^{-1-q}\left[1 + \frac{q[q+1]}{2j} + O(j^{-2})\right], \qquad j \to \infty,$$

1 INTRODUCTION

a representation which establishes that the $j \to \infty$ limit of $j^{q+1}\Gamma(j-q)/\Gamma(j+1)$ is unity. Thence one may show that

(1.3.14) $\quad \lim_{j \to \infty} \left[j^{c+q+1} \frac{\Gamma(j-q)}{\Gamma(j+1)} \right] = \lim_{j \to \infty} \left[j^{c+q} \frac{\Gamma(j-q)}{\Gamma(j)} \right] = \begin{cases} +\infty, & c > 0, \\ 1, & c = 0, \\ 0, & c < 0, \end{cases}$

a result that may be generalized to

(1.3.15) $\quad \lim_{j \to \infty} \left[j^{c+q+1} \frac{\Gamma(j+k-q)}{\Gamma(j+k+1)} \right] = \begin{cases} +\infty, & c > 0, \\ 1, & c = 0, \\ 0, & c < 0, \end{cases}$

for any finite integer k.

Expression (1.3.11) may be regarded as a binomial coefficient,

(1.3.16) $\quad \dfrac{\Gamma(j-q)}{\Gamma(-q)\Gamma(j+1)} = \binom{j-q-1}{j} = [-1]^j \binom{q}{j},$

the above equalities being readily established from the definition of a binomial coefficient and the reflection formula (1.3.8). Many relationships for gamma functions are derivable from the corresponding binomial coefficient identities [see Abramowitz and Stegun (1964, pp. 10, 822), as well as Gradshteyn and Ryzhik (1965, pp. 3, 4) for many of these]. Thus, since

(1.3.17) $\quad \displaystyle\sum_{j=0}^{n} \binom{j-q-1}{j} = \binom{n-q}{n},$

the relationship

(1.3.18) $\quad \displaystyle\sum_{j=0}^{N-1} \frac{\Gamma(j-q)}{\Gamma(-q)\Gamma(j+1)} = \frac{\Gamma(N-q)}{\Gamma(1-q)\Gamma(N)}$

follows on setting $N \equiv n+1$ and expressing the binomial coefficients as their equivalent gamma function combination. Similarly, after multiplication by $-q$, it is readily shown that

$$\sum_{j=1}^{n} \binom{j-q-1}{j-1} = \binom{n-q}{n-1}$$

[which follows straightforwardly from (1.3.17) on redefinition of q, j, and n] leads to the identity

(1.3.19) $\quad \displaystyle\sum_{j=0}^{N-1} \frac{\Gamma(j-q)}{\Gamma(-q)\Gamma(j)} = \frac{-q\Gamma(N-q)}{\Gamma(2-q)\Gamma(N-1)}.$

Likewise, the well-known summation formula

(1.3.20) $\quad \displaystyle\sum_{k=0}^{j} \binom{q}{k}\binom{Q}{j-k} = \binom{q+Q}{j}$

becomes the identity

(1.3.21)
$$\sum_{k=0}^{j} \frac{\Gamma(q+1)\Gamma(p+1)\Gamma(j+1)}{\Gamma(q-k+1)\Gamma(k+1)\Gamma(p-q+k+1)\Gamma(j-k+1)} = \frac{\Gamma(p+j+1)}{\Gamma(p-q+j+1)}$$

on setting Q equal to $[p-q+j]$ and multiplication by $\Gamma(j+1)$. Yet again, the binomial relationship

$$\sum_{j=0}^{n} \binom{j-q-1}{j-m} = \binom{n-q}{n-m},$$

which is obtainable from (1.3.17), leads easily to the gamma function equivalent of

(1.3.22)
$$\sum_{j=0}^{n} \binom{j-q-1}{j}\binom{j}{m} = \binom{m-q-1}{m}\binom{n-q}{n-m},$$

a relationship which we shall need in Section 3.5.

A function that is closely related to the gamma function is the complete beta function $B(p, q)$. For positive values of the two parameters, p and q, the function is defined by the beta integral,

(1.3.23) $$B(p, q) = \int_{0}^{1} y^{p-1}[1-y]^{q-1}\, dy, \qquad p > 0 < q,$$

also known as Euler's integral of the second kind. If either p or q is nonpositive, the integral diverges and the beta function is then defined by the relationship

(1.3.24) $$B(p, q) = \frac{\Gamma(p)\Gamma(q)}{\Gamma(p+q)},$$

valid for all p and q.

Both the beta function and the gamma function have "incomplete" analogs. The incomplete beta function of argument x is defined by the integral

(1.3.25) $$B_x(p, q) = \int_{0}^{x} y^{p-1}[1-y]^{q-1}\, dy.$$

There are many alternative formulations of the incomplete gamma function. The one which we shall adopt exclusively is that defined by

(1.3.26) $$\gamma^*(c, x) = \frac{c^{-x}}{\Gamma(x)} \int_{0}^{c} y^{x-1} \exp(-y)\, dy = \exp(-x) \sum_{j=0}^{\infty} \frac{x^j}{\Gamma(j+c+1)};$$

22   1   INTRODUCTION

$\gamma^*(c, x)$ is a finite single-valued analytic function of x and c. Properties of the incomplete gamma function of which we shall have need in Chapter 8 are its recursion

(1.3.27) $$\gamma^*(c - 1, x) = x\gamma^*(c, x) + \frac{\exp(-x)}{\Gamma(c)},$$

and its value

(1.3.28) $$\gamma^*(\tfrac{1}{2}, x) = \frac{\operatorname{erf}(\sqrt{x})}{\sqrt{x}}$$

for a parameter of moiety.

Powers of numbers may be expressed in terms of complete gamma functions, thus

$$q^j = \sum_{m=0}^{j} [-]^m S_j^{[m]} \frac{\Gamma(m - q)}{\Gamma(-q)}.$$

Table 1.3.3 *Some examples of Stirling numbers of the second kind, $S_j^{[m]}$*

j	\multicolumn{6}{c}{m}					
	0	1	2	3	4	5
0	1	0	0	0	0	0
1	0	1	0	0	0	0
2	0	1	1	0	0	0
3	0	1	3	1	0	0
4	0	1	7	6	1	0
5	0	1	15	25	10	1

The coefficients $S_j^{[m]}$ appearing in this expansion are Stirling numbers of the second kind. Examples of these numbers, which obey the recurrence

$$S_j^{[m]} = S_j^{[m-1]} + m S_j^{[m]}; \quad S_0^{[m]} = S_j^{[0]} = 0, \quad \text{except} \quad S_0^{[0]} = 1,$$

are listed in Table 1.3.3. Notice that

(1.3.29) $$S_j^{[j]} = 1$$

for all j, and that

(1.3.30) $$S_j^{[m]} = 0, \quad j = 0, 1, \ldots, m - 1.$$

The formula

(1.3.31) $$\sum_{l=0}^{m} [-]^{l+m} \binom{m}{l} l^k = m! \, S_k^{[m]}$$

may be regarded as defining Stirling numbers of the second kind.

1.3 PROPERTIES OF THE GAMMA FUNCTION

In Section 3.5 we shall need to expand the analytic function ϕ of argument $(x + jy)$ in a rather special way. We proceed to sketch the proof of this expansion, which involves Stirling numbers of the second kind. First, we relate $\phi(x + jy)$ to the values $\phi(x)$, $\phi(x + y)$, $\phi(x + 2y)$, ..., $\phi(x + jy)$ by the formula

$$\phi(x + jy) = \sum_{m=0}^{j} \binom{j}{m} \sum_{l=0}^{m} [-1]^{l+m} \binom{m}{l} \phi(x + ly).$$

The inner summation is now symbolized $G_m(\phi, x, y)$ and Taylor expanded to give

$$G_m(\phi, x, y) = \sum_{l=0}^{m} [-1]^{l+m} \binom{m}{l} \sum_{k=0}^{\infty} \frac{[ly]^k}{k!} \phi^{(k)}(x)$$

$$= \sum_{k=0}^{\infty} \frac{y^k}{k!} \phi^{(k)}(x) \sum_{l=0}^{m} [-1]^{l+m} \binom{m}{l} l^k.$$

It now only requires the application of summation (1.3.31) to yield the final result:

(1.3.32) $$\phi(x + jy) = \sum_{m=0}^{j} G_m(\phi, x, y) \binom{j}{m},$$

where

(1.3.33) $$G_m(\phi, x, y) = \sum_{k=0}^{\infty} \frac{m!}{k!} S_k^{[m]} y^k \phi^{(k)}(x).$$

Because $\binom{j}{m}$ vanishes when m exceeds the integer j, the upper summation limit in (1.3.32) may be replaced by ∞. Similarly, but this time as a consequence of equation (1.3.30), the $k = 0$ lower summation limit in expansion (1.3.33) may be replaced by $k = m$.

Yet another relative of the gamma function is the psi function that we shall come across in Section 6.7. Defined by

$$\psi(x) \equiv \frac{1}{\Gamma(x)} \frac{d\Gamma(x)}{dx},$$

the psi function obeys the recursion

$$\psi(x + 1) = \psi(x) + x^{-1},$$

whence it follows that

(1.3.34) $$\psi(n + 1) = \psi(1) + \sum_{j=1}^{n} \frac{1}{j},$$

where

(1.3.35) $$-\psi(1) = \gamma = 0.5772157 \cdots$$

is Euler's constant. Our encounter with $\psi(x)$ in Section 6.7 is via the definite integral

(1.3.36) $$\int_0^1 \frac{[v^x - v^y]\, dv}{1 - v} = \psi(y + 1) - \psi(x + 1).$$

CHAPTER 2
DIFFERENTIATION AND INTEGRATION TO INTEGER ORDER

Our point of view in this work is that many results about generalized derivatives and integrals are motivated by and follow rather naturally from classical results about derivatives and integrals of integer order. Accordingly, after first establishing some notational conventions, we shall present familiar definitions and a number of results for ordinary multiple derivatives and integrals. In this way we hope to motivate the corresponding definitions and results for derivatives and integrals of arbitrary order when they are encountered in Chapter 3.

2.1 SYMBOLISM

We are accustomed to the use of the notation

$$\frac{d^n f}{dx^n}$$

for the nth derivative of a function f with respect to x when n is a nonnegative integer. Because integration and differentiation are inverse operations,[1] it is natural to associate the symbolism

$$\frac{d^{-1} f}{[dx]^{-1}}$$

with indefinite integration of f with respect to x. However, it is necessary to stipulate a lower limit of integration in order that an indefinite integral be

---

[1] Except when they are not! See the discussion in Section 2.3 relating to classical differentiation and integration and the more general discussion in Section 5.7.

completely specified. We choose to associate the above symbol with the lower limit zero. Hence we define

$$\frac{d^{-1}f}{[dx]^{-1}} \equiv \int_0^x f(y)\, dy.$$

Multiple integration with zero lower limit is symbolized by the natural extensions

$$\frac{d^{-2}f}{[dx]^{-2}} \equiv \int_0^x dx_1 \int_0^{x_1} f(x_0)\, dx_0,$$

$$\vdots$$

$$\frac{d^{-n}f}{[dx]^{-n}} \equiv \int_0^x dx_{n-1} \int_0^{x_{n-1}} dx_{n-2} \cdots \int_0^{x_2} dx_1 \int_0^{x_1} f(x_0)\, dx_0.$$

Use is made of the identity

$$\int_a^x f(y)\, dy = \int_0^{x-a} f(y+a)\, dy$$

to extend the symbolism to lower limits other than zero. Thus we define

$$\frac{d^{-1}f}{[d(x-a)]^{-1}} \equiv \int_a^x f(y)\, dy,$$

$$\vdots$$

$$\frac{d^{-n}f}{[d(x-a)]^{-n}} \equiv \int_a^x dx_{n-1} \int_a^{x_{n-1}} dx_{n-2} \cdots \int_a^{x_2} dx_1 \int_a^{x_1} f(x_0)\, dx_0.$$

Caution must be exercised against subconsciously carrying over the equivalence

$$\frac{d^n}{[d(x-a)]^n} = \frac{d^n}{dx^n}$$

characteristic of a local operator to negative orders (or, as we shall see in Section 5.8, to fractional orders of either sign), because, of course,

$$\frac{d^{-n}}{[d(x-a)]^{-n}} \neq \frac{d^{-n}}{[dx]^{-n}}$$

in general. We use these new symbols because of the clumsiness of the conventional symbolism for multiple integration and in order to achieve a unified terminology which embraces both integration and differentiation. We prefer our symbolism to

$$p^{-n} \quad \text{and} \quad I^n$$

used in the literature (Riesz, 1949; Duff, 1956; Courant and Hilbert, 1962) for n-fold integration because these do not specify the lower limit, nor do they lend themselves readily to symbolize integrations with respect to a variable other than the independent variable, x, of the function under discussion.

The symbol $f^{(n)}$ finds frequent use as an abbreviation for $d^n f/[dx]^n$: it has the merit of brevity. Likewise we shall occasionally use $f^{(-n)}$ to symbolize an n-fold integration of f with respect to x, the lower limits being unspecified. That is,

$$f^{(-n)} \equiv \int_{a_n}^{x} dx_{n-1} \int_{a_{n-1}}^{x_{n-1}} dx_{n-2} \cdots \int_{a_2}^{x_2} dx_1 \int_{a_1}^{x_1} f(x_0) \, dx_0,$$

where $a_1, a_2, \ldots, a_n$ are completely arbitrary. However, when we write a difference such as $f^{(-n)}(x) - f^{(-n)}(a)$ we intend that the same lower limits $a_1, a_2, \ldots, a_n$ attach to each integral. Finally, we notice that our habit has been to write f rather than $f(x)$ and

$$\frac{d^{-1}f}{[d(x-a)]^{-1}}$$

rather than

$$\frac{d^{-1}f}{[d(x-a)]^{-1}}(x)$$

since f and $d^{-1}f/[d(x-a)]^{-1}$ are understood to be functions of the independent variable x. When we have need to specify the value of x at which a function is to be evaluated we shall adopt symbols such as

$$f(a) \quad \text{or} \quad \left[\frac{d^{-1}f}{[d(x-a)]^{-1}}\right]_{x=x_0}$$

for that purpose.

2.2 CONVENTIONAL DEFINITIONS

In this section we shall review standard definitions of repeated derivatives and integrals as limits of difference quotients and sums, respectively. Our aim is to embrace both kinds of operations within a single formula and thus to pave the way for the extension we seek to differintegral[2] operators to arbitrary order.

[2] This word seems to us a natural one to describe a class of operators that includes both differential and integral operators as special cases.

We begin with the familiar definition of the first derivative in terms of a backward difference

$$\frac{d^1 f}{dx^1} \equiv \frac{d}{dx} f(x) \equiv \lim_{\delta x \to 0} \{[\delta x]^{-1}[f(x) - f(x - \delta x)]\}.$$

Similarly,

$$\frac{d^2 f}{dx^2} \equiv \lim_{\delta x \to 0} \{[\delta x]^{-2}[f(x) - 2f(x - \delta x) + f(x - 2\delta x)]\}$$

and

$$\frac{d^3 f}{dx^3} \equiv \lim_{\delta x \to 0} \{[\delta x]^{-3}[f(x) - 3f(x - \delta x) + 3f(x - 2\delta x) - f(x - 3\delta x)]\},$$

etc., where we have, of course, assumed that the indicated limits exist.

Observe that each derivative involves one more functional evaluation than the order of the derivative, and that the coefficients build up as binomial coefficients and alternate in sign. This suggests the general formula for positive integer n,

$$\frac{d^n f}{dx^n} \equiv \lim_{\delta x \to 0} \left\{ [\delta x]^{-n} \sum_{j=0}^{n} [-1]^j \binom{n}{j} f(x - j\,\delta x) \right\}.$$

If the nth derivative of f exists, this last equation does indeed define $d^n f/dx^n$ as an unrestricted limit, i.e., as a limit as δx tends to zero through values that are totally unrestricted. In order to unify this formula with the one which defines an integral as a limit of a sum, it is desirable to define derivatives in terms of a restricted limit, namely, as a limit as δx tends to zero through discrete values only. To do this, choose $\delta_N x \equiv [x - a]/N$, $N = 1, 2, \ldots$, where a is a number smaller than x and plays a role akin to a lower limit. Then, since if the unrestricted limit exists so does the restricted limit and they are equal, the nth derivative may be defined as

$$\frac{d^n f}{[dx]^n} \equiv \lim_{\delta_N x \to 0} \left\{ [\delta_N x]^{-n} \sum_{j=0}^{n} [-1]^j \binom{n}{j} f(x - j\,\delta_N x) \right\}.$$

Now, since $\binom{n}{j} = 0$ if $j > n$ when n is integer, the above may be rewritten

(2.2.1)
$$\frac{d^n f}{[dx]^n} \equiv \lim_{\delta_N x \to 0} \left\{ [\delta_N x]^{-n} \sum_{j=0}^{N-1} [-1]^j \binom{n}{j} f(x - j\,\delta_N x) \right\}$$

$$\equiv \lim_{N \to \infty} \left\{ \left[\frac{x-a}{N}\right]^{-n} \sum_{j=0}^{N-1} [-1]^j \binom{n}{j} f\!\left(x - j\left[\frac{x-a}{N}\right]\right) \right\}.$$

Equation (2.2.1) will henceforth be adopted as defining $d^n f/[dx]^n$ with the understanding that the limit exists in the usual, unrestricted sense.

2.2 CONVENTIONAL DEFINITIONS

Turning our attention from derivatives to integrals, we begin with the usual definition of an integral as a limit of a Riemann sum. Using the symbolism of Section 2.1,

$$\frac{d^{-1}f}{[d(x-a)]^{-1}} \equiv \int_a^x f(y)\,dy$$

$$\equiv \lim_{\delta_N x \to 0} \{\delta_N x[f(x) + f(x - \delta_N x) + f(x - 2\delta_N x) + \cdots$$

$$+ f(a + \delta_N x)]\}$$

$$\equiv \lim_{\delta_N x \to 0} \left\{ \delta_N x \sum_{j=0}^{N-1} f(x - j\delta_N x) \right\},$$

where $\delta_N x \equiv [x - a]/N$, as before. Application of the same definition to a double integral gives

$$\frac{d^{-2}f}{[d(x-a)]^{-2}} \equiv \int_a^x dx_1 \int_a^{x_1} f(x_0)\,dx_0$$

$$\equiv \lim_{\delta_N x \to 0} \{[\delta_N x]^2[f(x) + 2f(x - \delta_N x) + 3f(x - 2\delta_N x) + \cdots$$

$$+ Nf(a + \delta_N x)]\}$$

$$\equiv \lim_{\delta_N x \to 0} \left\{ [\delta_N x]^2 \sum_{j=0}^{N-1} [j+1] f(x - j\delta_N x) \right\}.$$

We need to do one more iteration to obtain a clearer picture of the general formula:

$$\frac{d^{-3}f}{[d(x-a)]^{-3}} \equiv \int_a^x dx_2 \int_a^{x_2} dx_1 \int_a^{x_1} f(x_0)\,dx_0$$

$$\equiv \lim_{\delta_N x \to 0} \left\{ [\delta_N x]^3 \sum_{j=0}^{N-1} \frac{[j+1][j+2]}{2} f(x - j\delta_N x) \right\}.$$

This time we notice the coefficients building up as $\binom{j+n-1}{j}$, where n is the order of the integral, and all the signs are positive. Therefore,

(2.2.2) $$\frac{d^{-n}f}{[d(x-a)]^{-n}} \equiv \lim_{\delta_N x \to 0} \left\{ [\delta_N x]^n \sum_{j=0}^{N-1} \binom{j+n-1}{j} f(x - j\delta_N x) \right\}$$

$$\equiv \lim_{N \to \infty} \left\{ \left[\frac{x-a}{N}\right]^n \sum_{j=0}^{N} \binom{j+n-1}{j} f\left(x - j\left[\frac{x-a}{N}\right]\right) \right\}.$$

We now compare formulas (2.2.2) and (2.2.1), recalling equation (1.3.16),

$$[-1]^j \binom{n}{j} = \binom{j-n-1}{j} = \frac{\Gamma(j-n)}{\Gamma(-n)\Gamma(j+1)},$$

and see that formulas (2.2.2) and (2.2.1) are embraced in the equation

$$(2.2.3) \quad \frac{d^q f}{[d(x-a)]^q} \equiv \lim_{N \to \infty} \left\{ \frac{\left[\frac{x-a}{N}\right]^{-q}}{\Gamma(-q)} \sum_{j=0}^{N-1} \frac{\Gamma(j-q)}{\Gamma(j+1)} f\left(x - j\left[\frac{x-a}{N}\right]\right) \right\},$$

where q is an integer of either sign. We shall have more to say about this formula later when we introduce it as our most basic definition of a differintegral.

2.3 COMPOSITION RULE FOR MIXED INTEGER ORDERS

The identities

$$(2.3.1) \quad \frac{d^n}{[dx]^n}\left\{\frac{d^N f}{[dx]^N}\right\} = \frac{d^{n+N} f}{[dx]^{n+N}} = \frac{d^N}{[dx]^N}\left\{\frac{d^n f}{[dx]^n}\right\}$$

and

$$(2.3.2) \quad \frac{d^{-n}}{[d(x-a)]^{-n}}\left\{\frac{d^{-N} f}{[d(x-a)]^{-N}}\right\} = \frac{d^{-n-N} f}{[d(x-a)]^{-n-N}} = \frac{d^{-N}}{[d(x-a)]^{-N}}\left\{\frac{d^{-n} f}{[d(x-a)]^{-n}}\right\}$$

are obeyed when n and N are nonnegative integers; indeed these identities are basic to the concepts of multiple differentiation and integration. This section seeks to investigate the conditions, if any, required to identify

$$(2.3.3) \quad \frac{d^{-n}}{[d(x-a)]^{-n}}\left\{\frac{d^N f}{[d(x-a)]^N}\right\},$$

$$(2.3.4) \quad \frac{d^{N-n} f}{[d(x-a)]^{N-n}},$$

and

$$(2.3.5) \quad \frac{d^N}{[d(x-a)]^N}\left\{\frac{d^{-n} f}{[d(x-a)]^{-n}}\right\},$$

and to evaluate the difference when any two are unequal.

Let f be a function which is at least N-fold differentiable [$N \geq 1$]; then by formula (2.3.1) and the fundamental theorem of the calculus

$$\frac{d^{-1} f^{(N)}}{[d(x-a)]^{-1}} = f^{(N-1)}(x) - f^{(N-1)}(a),$$

2.3 COMPOSITION RULE FOR MIXED INTEGER ORDERS

with the understanding that $f^{(0)}(x) \equiv f(x)$. Next we integrate from a to x a second time. Making use of equation (2.3.2), the left-hand side becomes $d^{-2}f^{(N)}/[d(x-a)]^{-2}$. There are, however, two cases to consider in treating the right-hand side. When $N \geq 2$, repetition of the use of formula (2.3.1), the fundamental theorem of the calculus and the rule for integrating a constant, leads to

$$\frac{d^{-2}f^{(N)}}{[d(x-a)]^{-2}} = f^{(N-2)}(x) - f^{(N-2)}(a) - [x-a]f^{(N-1)}(a).$$

When $N = 1$, this same result is obtained utilizing the identity

$$(2.3.6) \qquad \frac{d^{-1}f}{[d(x-a)]^{-1}} = \int_a^x f(y)\,dy = f^{(-1)}(x) - f^{(-1)}(a).$$

The relationship

$$(2.3.7) \qquad \frac{d^{-n}f^{(N)}}{[d(x-a)]^{-n}} = f^{(N-n)}(x) - \sum_{k=0}^{n-1} \frac{[x-a]^k}{k!} f^{(N+k-n)}(a)$$

follows from n repetitions of this procedure.[3]

Notice that equation (2.3.7) becomes

$$f(x) = \sum_{k=0}^{n-1} \frac{[x-a]^k}{k!} f^{(k)}(a) + R_n$$

after setting $N = n$ and rearrangement. This is Taylor's formula in which the remainder

$$R_n = \frac{d^{-n}f^{(n)}}{[d(x-a)]^{-n}}$$

is expressed as the n-fold integral of the n-fold derivative of f.

Upon repeated integration of equation (2.3.6), we obtain

$$(2.3.8) \qquad \frac{d^{-n}f}{[d(x-a)]^{-n}} = f^{(-n)}(x) - \sum_{k=0}^{n-1} \frac{[x-a]^k}{k!} f^{(k-n)}(a).$$

Differentiation, using Leibniz's theorem (Abramowitz and Stegun, 1964, p. 11) for differentiating an integral, gives

$$\frac{d}{d(x-a)}\left\{\frac{d^{-n}f}{[d(x-a)]^{-n}}\right\} = f^{(1-n)}(x) - \sum_{k=1}^{n-1} \frac{[x-a]^{k-1}}{(k-1)!} f^{(k-n)}(a),$$

---

[3] By setting $N = n$ in equation (2.3.7), we see that the operators d^n and d^{-n} are not inverse to each other unless the function f as well as its first $n-1$ derivatives vanish at $x = a$. (Recall footnote 1 in this chapter.)

and after N such differentiations the equation

(2.3.9) $$\frac{d^N}{[d(x-a)]^N}\left\{\frac{d^{-n}f}{[d(x-a)]^{-n}}\right\} = f^{(N-n)}(x) - \sum_{k=N}^{n-1}\frac{[x-a]^{k-N}}{(k-N)!}f^{(k-n)}(a)$$

emerges.[4]

If $N \geqq n$, we have

(2.3.10) $$\frac{d^{N-n}f}{[d(x-a)]^{N-n}} = f^{(N-n)}$$

because the choice of lower limit does not affect differentiation. However, if $N < n$, by analogy with equation (2.3.8),

(2.3.11) $$\frac{d^{N-n}f}{[d(x-a)]^{N-n}} = f^{(N-n)}(x) - \sum_{k=0}^{n-N-1}\frac{[x-a]^k}{k!}f^{(k+N-n)}(a),$$

a result which encompasses equation (2.3.10). After a redefinition of the summation index in expression (2.3.9), we see that the right-hand sides and, therefore, the left-hand sides of equations (2.3.9) and (2.3.11) are identical.

Summarizing, we have demonstrated that

(2.3.12) $$\frac{d^N}{[d(x-a)]^N}\left\{\frac{d^{-n}f}{[d(x-a)]^{-n}}\right\} = \frac{d^{N-n}f}{[d(x-a)]^{N-n}}$$

$$= \frac{d^{-n}}{[d(x-a)]^{-n}}\left\{\frac{d^N f}{[d(x-a)]^N}\right\}$$

$$+ \sum_{k=n-N}^{n-1}\frac{[x-a]^k}{k!}f^{(k+N-n)}(a).$$

That is, the composition "rule"

$$\frac{d^q}{[d(x-a)]^q}\left\{\frac{d^Q f}{[d(x-a)]^Q}\right\} = \frac{d^{q+Q}f}{[d(x-a)]^{q+Q}}$$

(when q and Q are integers) necessarily holds unless Q is positive and q is negative; in words, unless f is first differentiated and then integrated. The identity of expression (2.3.3) with the other two will hold only if $f(a) = 0$ and if all derivatives of f through the $(N-1)$th are also zero at $x = a$.

As an example of the failure of the equality for mixed signs, consider $f = \exp(2x) + 1$ with $N = 1$, $n = 3$, and $a = 0$. Then

(2.3.13) $$\frac{d}{dx}\left\{\frac{d^{-3}}{[dx]^{-3}}[\exp(2x)+1]\right\} = \frac{d^{-2}}{[dx]^{-2}}[\exp(2x)+1]$$

$$= \tfrac{1}{4}\exp(2x) + \tfrac{1}{2}x^2 - \tfrac{1}{2}x - \tfrac{1}{4},$$

---

[4] Notice that the summation is empty if $N \geqq n$.

whereas

(2.3.14) $$\frac{d^{-3}}{[dx]^{-3}}\left\{\frac{d}{dx}[\exp(2x)+1]\right\} = \tfrac{1}{4}\exp(2x) - \tfrac{1}{2}x^2 - \tfrac{1}{2}x - \tfrac{1}{4}.$$

The difference x^2 between expressions (2.3.13) and (2.3.14) is correctly given by formula (2.3.12) as

$$\sum_{k=2}^{2}\frac{x^k}{k!}f^{(k-2)}(0).$$

2.4 DEPENDENCE OF MULTIPLE INTEGRALS ON LOWER LIMIT

A multiple indefinite integral depends in a rather complex fashion on the magnitude of its lower limit. Thus, making use of equation (2.3.8),

(2.4.1)
$$\frac{d^{-n}f}{[d(x-a)]^{-n}} - \frac{d^{-n}f}{[d(x-b)]^{-n}} = \sum_{k=0}^{n-1}\frac{1}{k!}\{[x-b]^k f^{(k-n)}(b) - [x-a]^k f^{(k-n)}(a)\}.$$

Apart from the exceptional $n = 1$ case,

$$\frac{d^{-1}f}{[d(x-a)]^{-1}} - \frac{d^{-1}f}{[d(x-b)]^{-1}} = f^{(-1)}(b) - f^{(-1)}(a),$$

the difference in formula (2.4.1) depends not only on the lower limits, but also on the upper limit x.

If we seek the explicit dependence on $b - a$ of shifting the lower limit from b to a, we start with equation (2.4.1) and proceed as follows:

$$\frac{d^{-n}f}{[d(x-a)]^{-n}} - \frac{d^{-n}f}{[d(x-b)]^{-n}} - \sum_{k=0}^{n-1}\frac{[x-b]^k}{k!}f^{(k-n)}(b)$$

$$= -\sum_{k=0}^{n-1}\frac{[(x-b)+(b-a)]^k}{k!}f^{(k-n)}(a)$$

$$= -\sum_{k=0}^{n-1}f^{(k-n)}(a)\sum_{j=0}^{k}\frac{[x-b]^j[b-a]^{k-j}}{j!(k-j)!}$$

$$= -\sum_{K=0}^{n-1}\frac{[x-b]^K}{K!}\sum_{J=K}^{n-1}\frac{[b-a]^{J-K}}{(J-K)!}f^{(J-n)}(a)$$

$$= -\sum_{K=0}^{n-1}\frac{[x-b]^K}{K!}\sum_{j=0}^{n-K-1}\frac{[b-a]^j}{j!}f^{(j+K-n)}(a),$$

where there have been many changes of summation index. Identifying k with K and later replacing $n - K$ by k, we find

$$\frac{d^{-n}f}{[d(x-a)]^{-n}} - \frac{d^{-n}f}{[d(x-b)]^{-n}}$$
$$= \sum_{K=0}^{n-1} \frac{[x-b]^K}{K!} \left[f^{(K-n)}(b) - \sum_{j=0}^{n-K-1} \frac{[b-a]^j}{j!} f^{(K+j-n)}(a) \right]$$
$$= \sum_{k=1}^{n} \frac{[x-b]^{n-k}}{(n-k)!} \left[f^{(-k)}(b) - \sum_{j=0}^{k-1} \frac{[b-a]^j}{j!} f^{(j-k)}(a) \right]$$
$$= \sum_{k=1}^{n} \frac{[x-b]^{n-k}}{(n-k)!} \frac{d^{-k}f(b)}{[d(b-a)]^{-k}},$$

where equation (2.3.8) was used once again for the final step.

2.5 PRODUCT RULE FOR MULTIPLE INTEGRALS

We are interested in establishing a rule for multiple integration of a product of two functions, similar to Leibniz's theorem for repeatedly differentiating a product. In Section 5.5 a more general result will be established that embraces both Leibniz's theorem and the result of this section.

We begin with the familiar formula for integration by parts:

(2.5.1) $\int_a^x g(y)\, dv(y) = g(x)v(x) - g(a)v(a) - \int_a^x v(y)\, dg(y).$

Let

$$v(y) = \int_a^y f(z)\, dz$$

in (2.5.1); then

$$\int_a^x g(y)f(y)\, dy = g(x) \int_a^x f(z)\, dz - \int_a^x \left[\int_a^y f(z)\, dz \right] \frac{dg(y)}{dy}\, dy$$

or, in the symbolism of Section 2.1,

(2.5.2) $\dfrac{d^{-1}[fg]}{[d(x-a)]^{-1}} = g\, \dfrac{d^{-1}f}{[d(x-a)]^{-1}} - \dfrac{d^{-1}}{[d(x-a)]^{-1}} \left\{ g^{(1)}\, \dfrac{d^{-1}f}{[d(x-a)]^{-1}} \right\}.$

2.5 PRODUCT RULE FOR MULTIPLE INTEGRALS

When (2.5.2) is applied to the product in braces, and the composition rule (2.3.2) evoked, one has

$$\frac{d^{-1}[fg]}{[d(x-a)]^{-1}} = g\frac{d^{-1}f}{[d(x-a)]^{-1}} - g^{(1)}\frac{d^{-2}f}{[d(x-a)]^{-2}}$$
$$+ \frac{d^{-1}}{[d(x-a)]^{-1}}\left\{g^{(2)}\frac{d^{-2}f}{[d(x-a)]^{-2}}\right\}$$

and indefinite repetition of this process gives [utilizing equation (1.3.16)]

(2.5.3)
$$\frac{d^{-1}[fg]}{[d(x-a)]^{-1}} = \sum_{j=0}^{\infty}[-]^{j}g^{(j)}\frac{d^{-1-j}f}{[d(x-a)]^{-1-j}}$$
$$= \sum_{j=0}^{\infty}\binom{-1}{j}g^{(j)}\frac{d^{-1-j}f}{[d(x-a)]^{-1-j}}.$$

When (2.5.3) is integrated, using this same formula inside the summation, and the composition rules for integrals and derivatives (2.3.2) and (2.3.1) are applied, we obtain

(2.5.4)
$$\frac{d^{-2}[fg]}{[d(x-a)]^{-2}} = \sum_{j=0}^{\infty}\binom{-1}{j}\sum_{k=0}^{\infty}\binom{-1}{k}g^{(j+k)}\frac{d^{-j-k-2}f}{[d(x-a)]^{-j-k-2}}$$
$$= \sum_{j=0}^{\infty}\sum_{l=j}^{\infty}\binom{-1}{j}\binom{-1}{l-j}g^{(l)}\frac{d^{-2-l}f}{[d(x-a)]^{-2-l}}$$
$$= \sum_{l=0}^{\infty}\sum_{j=0}^{l}\binom{-1}{j}\binom{-1}{l-j}g^{(l)}\frac{d^{-2-l}f}{[d(x-a)]^{-2-l}}$$
$$= \sum_{l=0}^{\infty}\binom{-2}{l}g^{(l)}\frac{d^{-2-l}f}{[d(x-a)]^{-2-l}}.$$

In the penultimate and final steps of (2.5.4), we have made use of the permutation[5]

$$\sum_{k=0}^{\infty}\sum_{j=0}^{k} = \sum_{j=0}^{\infty}\sum_{k=j}^{\infty}$$

and of the equation (1.3.20).

Iteration of the procedure that produced (2.5.4) from (2.5.3) leads to the desired formula:

$$\frac{d^{-n}[fg]}{[d(x-a)]^{-n}} = \sum_{j=0}^{\infty}\binom{-n}{j}g^{(j)}\frac{d^{-n-j}f}{[d(x-a)]^{-n-j}}, \qquad n = 1, 2, 3, \ldots.$$

[5] This identity is easily established, much as is the analogous permutation of variables in the double integral $\int_0^\infty dy \int_0^y dx = \int_0^\infty dx \int_x^\infty dy$.

2.6 THE CHAIN RULE FOR MULTIPLE DERIVATIVES

The chain rule for differentiation,

(2.6.1) $$\frac{d}{dx}g(f(x)) = \frac{d}{du}g(u)\frac{d}{dx}f(x) = g^{(1)}f^{(1)},$$

which enables $g(u)$ to be differentiated with respect to x if the derivatives of $g(u)$ with respect to u and of u with respect to x are known, is one of the most useful in the differential calculus. In this section we report the extension of this rule to higher orders of differentiation. For brevity, we shall use $g^{(m)}$ and $f^{(n)}$ to denote

$$\frac{d^m}{du^m}g(u) \quad \text{and} \quad \frac{d^n}{dx^n}f(x),$$

respectively.

By application of (2.6.1) and Leibniz's rule for differentiating a product, we find

$$\frac{d^2}{dx^2}g(f(x)) = \frac{d}{dx}\left[\frac{d}{dx}g(f(x))\right]$$

$$= \frac{d}{dx}\left[\frac{d}{du}g(u)\frac{d}{dx}f(x)\right]$$

$$= \left[\frac{d}{dx}\frac{d}{du}g(u)\right]\left[\frac{d}{dx}f(x)\right] + \left[\frac{d}{du}g(u)\right]\left[\frac{d^2}{dx^2}f(x)\right]$$

$$= \left[\frac{du}{dx}\frac{d^2}{du^2}g(u)\right]\left[\frac{d}{dx}f(x)\right] + \left[\frac{d}{du}g(u)\right]\left[\frac{d^2}{dx^2}f(x)\right]$$

$$= g^{(1)}f^{(2)} + g^{(2)}[f^{(1)}]^2.$$

The repetition of similar procedures yields successively

$$\frac{d^3}{dx^3}g(f(x)) = g^{(1)}f^{(3)} + 3g^{(2)}f^{(1)}f^{(2)} + g^{(3)}[f^{(1)}]^3,$$

$$\frac{d^4}{dx^4}g(f(x)) = g^{(1)}f^{(4)} + 4g^{(2)}f^{(1)}f^{(3)} + 6g^{(2)}[f^{(2)}]^2$$
$$+ 6g^{(3)}[f^{(1)}]^2 f^{(2)} + g^{(4)}[f^{(1)}]^4,$$

$$\frac{d^5}{dx^5}g(f(x)) = g^{(1)}f^{(5)} + 5g^{(2)}f^{(1)}f^{(4)} + 10g^{(3)}[f^{(1)}]^2 f^{(3)}$$
$$+ 30g^{(3)}f^{(1)}[f^{(2)}]^2 + 10g^{(4)}[f^{(1)}]^3 f^{(2)} + g^{(5)}[f^{(1)}]^5,$$

etc. The generalization to order n produces Faà di Bruno's formula (Abramowitz and Stegun, 1964, p. 823)

$$\frac{d^n}{dx^n} g(f(x)) = n! \sum_{m=1}^{n} g^{(m)} \sum \prod_{k=1}^{n} \frac{1}{P_k!} \left[\frac{f^{(k)}}{k!}\right]^{P_k},$$

where $\sum$ extends over all combinations of nonnegative integer values of $P_1, P_2, \ldots, P_n$ such that

$$\sum_{k=1}^{n} kP_k = n \quad \text{and} \quad \sum_{k=1}^{n} P_k = m.$$

Faà di Bruno's formula is sufficiently complicated to be of little general utility for large n. However, certain specific instances of f and g reveal its power. Thus when $f = \exp(x)$,

$$\frac{d^n}{dx^n} g(\exp(x)) = \exp(nx) \sum_{m=1}^{n} S_n^{[m]} g^{(m)},$$

where $S_n^{[m]}$ is a Stirling number of the second kind (see Section 1.3). Again, if $g = u^2$, we find

$$\frac{d^n}{dx^n} [f]^2 = 2f^{(n)}f + \sum_{k=1}^{n-1} \binom{n}{k} f^{(k)} f^{(n-k)},$$

a result that may also be obtained from Leibniz's theorem for the multiple differentiation of a product.

2.7 ITERATED INTEGRALS

Consider the formula

$$(2.7.1) \quad \frac{d^{-1}f}{[d(x-a)]^{-1}} \equiv \int_a^x f(y)\, dy = \frac{1}{n!} \frac{d^n}{dx^n} \int_a^x [x-y]^n f(y)\, dy,$$

$$n = 0, 1, 2, \ldots.$$

For $n = 0$, (2.7.1) is an identity, while for $n = 1$ it follows easily from Leibniz's theorem for differentiating an integral (Abramowitz and Stegun, 1964, p. 11). For general integer n one need only notice that the evaluation of the integrand on the right-hand side at the upper limit x gives 0, while differentiation n times inside the integral produces $n!f(y)$.

A single integration of (2.7.1) for $n = 1$ produces

$$\frac{d^{-2}f}{[d(x-a)]^{-2}} \equiv \int_a^x dx_1 \int_a^{x_1} f(x_0)\, dx_0 = \frac{1}{1!} \int_a^x [x-y] f(y)\, dy,$$

and an $[n-1]$-fold integration produces Cauchy's formula for repeated integration:

(2.7.2)
$$\frac{d^{-n}f}{[d(x-a)]^{-n}} \equiv \int_a^x dx_{n-1} \int_a^{x_{n-1}} \cdots \int_a^{x_1} f(x_0)\, dx_0 = \frac{1}{(n-1)!} \int_a^x [x-y]^{n-1} f(y)\, dy.$$

Thus an iterated integral may be expressed as a weighted single integral with a very simple weight function, a fact that provides an important clue for generalizations involving noninteger orders.

2.8 DIFFERENTIATION AND INTEGRATION OF SERIES

A great many functions are traditionally described by infinite series expansions. It is of utmost importance, therefore, to understand the conditions that permit term-by-term differentiation or integration of such infinite series. We recall here the two classical results that apply; Section 5.2 gives extensions of these classical results to differintegrals of arbitrary order.

Suppose $f_0, f_1, \ldots$ are functions defined and continuous on the closed interval $a \leq x \leq b$. Then

(2.8.1)
$$\frac{d^{-1}}{[d(x-a)]^{-1}} \left\{ \sum_{j=0}^{\infty} f_j \right\} = \sum_{j=0}^{\infty} \frac{d^{-1} f_j}{[d(x-a)]^{-1}}, \qquad a \leq x \leq b,$$

provided the series $\sum f_j$ converges uniformly in the interval $a \leq x \leq b$.

The hypotheses required to distribute a derivative through the terms of an infinite series are somewhat different. For this we need each f_j to have continuous derivatives on $a \leq x \leq b$. Then

(2.8.2)
$$\frac{d}{dx} \left\{ \sum_{j=0}^{\infty} f_j \right\} = \sum_{j=0}^{\infty} \frac{df_j}{dx}, \qquad a \leq x \leq b,$$

provided $\sum f_j$ converges pointwise and $\sum df_j/dx$ converges uniformly on the interval $a \leq x \leq b$.

Thus we see that a uniformly convergent series of continuous functions (which itself defines a continuous function) may be integrated term by term, and that a continuous series of continuously differentiable functions may be differentiated term by term provided that the derived series is uniformly convergent. The reader is referred to the work of Widder (1947) for proofs and further discussion of these theorems.

2.9 DIFFERENTIATION AND INTEGRATION OF POWERS

The previous section dealt with differentiation and integration of infinite series of functions. Throughout much of the rest of the book we shall often encounter series whose general term is the simple power $[x - a]^p$. We collect here the elementary formulas that express $d^q[x - a]^p/[d(x - a)]^q$ for positive and negative integer values of q. We have

(2.9.1) $$\frac{d^n[x - a]^p}{dx^n} = p[p - 1] \cdots [p - n + 1][x - a]^{p-n}, \qquad n = 0, 1, \ldots$$

and

(2.9.2) $$\frac{d^{-n}[x - a]^p}{[d(x - a)]^{-n}} \equiv \int_a^x dx_{n-1} \int_a^{x_{n-1}} dx_{n-2} \cdots \int_a^{x_1} [x_0 - a]^p \, dx_0$$

$$= \begin{cases} \dfrac{[x - a]^{p+n}}{[p + 1][p + 2] \cdots [p + n]}, & p > -1, \\ \infty, & p \leqq -1, \end{cases}$$

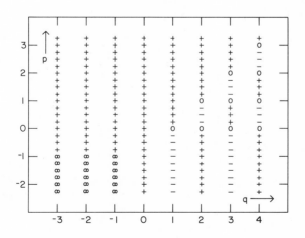

FIG. 2.9.1 The sign of $d^q[x - a]^p/[d(x - a)]^q$ for positive and negative integer values of q.

$n = 1, 2, \ldots$. Invoking the properties of the gamma function (see Section 1.3) and recalling that $d^n/dx^n = d^n/[d(x-a)]^n$, both of these formulas may be unified as

$$(2.9.3) \quad \frac{d^q[x-a]^p}{[d(x-a)]^q} = \begin{cases} \dfrac{\Gamma(p+1)[x-a]^{p-q}}{\Gamma(p-q+1)} & \begin{cases} q = 0, 1, \ldots; & \text{all } p, \\ q = -1, -2, \ldots; & p > -1, \end{cases} \\ \infty, & q = -1, -2, \ldots; \quad p \leq -1. \end{cases}$$

The coefficient $\Gamma(p+1)/\Gamma(p-q+1)$ may be positive, negative, or zero, according to the values of p and q. Figure 2.9.1 displays these instances graphically. Chapter 4 will extend formula (2.9.3) to noninteger q values, at which time the utility of the information conveyed by Fig. 2.9.1 should become more apparent.

2.10 DIFFERENTIATION AND INTEGRATION OF HYPERGEOMETRICS

In this section we discuss a generalized hypergeometric function and develop some of its properties, including its behavior under multiple differentiation and integration. This function will prove invaluable in Chapter 9 and is also considered in Section 6.6.

We adopt the abbreviated symbolism

$$(2.10.1) \quad \left[x \frac{b_1, b_2, \ldots, b_K}{c_1, c_2, \ldots, c_L} \right] = \sum_{j=0}^{\infty} x^j \frac{\prod_{k=1}^{K} \Gamma(j+1+b_k)}{\prod_{l=1}^{L} \Gamma(j+1+c_l)}$$

for what we shall term a $\frac{K}{L}$ hypergeometric. Note that leading terms within this summation will be zero if one of the denominatorial parameters is a negative integer: if $c_l = -n$, terms in $x^0, x^1, x^2, \ldots, x^{n-1}$ are absent. Conversely, infinite leading terms will be present if one of the numeratorial parameters, say b_1, is a negative integer; we consequently encounter this possibility only as the quotient

$$\frac{1}{\Gamma(1-n)} \left[x \frac{-n, b_2, \ldots, b_K}{c_1, c_2, \ldots, c_L} \right]$$

[see the discussion in Section 1.3 surrounding equation (1.3.4)] which represents a polynomial of degree $n - 1$. We assume that parameter values and the value of the argument are such as to ensure convergence of the series; this will often imply the restriction $|x| < 1$. Representation (2.10.1) proves to be more convenient than the usual symbol (Gradshteyn and Ryzhik, 1965, p. 1045) for generalized hypergeometric functions; the equation

$$\left[x \frac{b_1, b_2, \ldots, b_K}{c_1, c_2, \ldots, c_L} \right]$$

$$= {}_{K+1}F_L(1, 1 + b_1, 1 + b_2, \ldots, 1 + b_K; 1 + c_1, \ldots, 1 + c_L; x) \frac{\prod_{k=1}^{K} \Gamma(1 + b_k)}{\prod_{l=1}^{L} \Gamma(1 + c_l)}$$

relates the two. Some familiar functions which are instances of such hypergeometrics include

$$\left[x \frac{}{} \right] = \frac{1}{1 - x},$$

$$\left[x \frac{}{0} \right] = \exp(x),$$

$$\left[x \frac{}{c} \right] = \exp(x)\gamma^*(c, x),$$

$$\left[x \frac{b}{0} \right] = \frac{\Gamma(1 + b)}{[1 - x]^{1+b}},$$

$$\left[x \frac{b}{c} \right] = \frac{\Gamma(1 + b)}{\Gamma(c)} \frac{B_x(c, 1 + b - c)}{x^c [1 - x]^{1+b-c}},$$

$$\left[x \frac{b}{0, c} \right] = \frac{\Gamma(1 + b)}{\Gamma(1 + c)} M(1 + b, 1 + c, x),$$

and

$$\left[x \frac{b_1, b_2}{0, c} \right] = \frac{\Gamma(1 + b_1)\Gamma(1 + b_2)}{\Gamma(1 + c)} F(1 + b_1, 1 + b_2; 1 + c; x),$$

where $\gamma^*(\,, x)$, $B_x(\,,)$, $M(\,, , x)$, and $F(\,, ; ; x)$ denote the incomplete gamma, incomplete beta, Kummer, and Gauss functions of x, as defined by Abramowitz and Stegun (1964, Chapters 6, 13, 15).

The relationships

$$\left[x\frac{b_1+1, b_2+1, \ldots, b_K+1}{c_1+1, c_2+1, \ldots, c_L+1}\right] = \frac{1}{x}\left[x\frac{b_1, b_2, \ldots, b_K}{c_1, c_2, \ldots, c_L}\right]$$

$$-\frac{\Gamma(b_1+1)\Gamma(b_2+1)\cdots\Gamma(b_K+1)}{x\Gamma(c_1+1)\Gamma(c_2+1)\cdots\Gamma(c_L+1)}$$

and

$$\left[x\frac{b_1-1, b_2-1, \ldots, b_K-1}{c_1-1, c_2-1, \ldots, c_L-1}\right] = x\left[x\frac{b_1, b_2, \ldots, b_K}{c_1, c_2, \ldots, c_L}\right]$$

$$+\frac{\Gamma(b_1)\Gamma(b_2)\cdots\Gamma(b_K)}{\Gamma(c_1)\Gamma(c_2)\cdots\Gamma(c_L)}$$

are immediate consequences of definition (2.10.1). These formulas permit *all* the parameters to be increased or decreased by unity, but recurrences such as

$$\left[x\frac{b+1, b_2, \ldots, b_K}{c+1, c_2, \ldots, c_L}\right] = \left[x\frac{b, b_2, \ldots, b_K}{c, c_2, \ldots, c_L}\right]$$

$$+ [b-c]\left[x\frac{b, b_2, \ldots, b_K}{c+1, c_2, \ldots, c_L}\right]$$

and

$$\left[x\frac{b-1, b_2, \ldots, b_K}{c-1, c_2, \ldots, c_L}\right] = \left[x\frac{b, b_2, \ldots, b_K}{c, c_2, \ldots, c_L}\right]$$

$$+ [c-b]\left[x\frac{b, b_2, \ldots, b_K}{c-1, c_2, \ldots, c_L}\right]$$

enable single parameters to be selectively incremented or decremented. If one of the numeratorial parameters equals one of the denominatorial parameters, they may be "cancelled" as in the example

$$\left[x\frac{b}{b, c}\right] = \left[x\frac{}{c}\right],$$

2.10 DIFFERENTIATION AND INTEGRATION OF HYPERGEOMETRICS

thereby reducing the complexity of the hypergeometric function from $\frac{K}{L}$ to $\frac{K-1}{L-1}$.

The rule for multiple differentiation or integration of a generalized hypergeometric function is readily derived from the defining formula (2.10.1) because the differentiation or integration operator may be distributed through the infinite sum of terms. The rule is

$$(2.10.2) \quad \frac{d^q}{dx^q}\left[x \frac{b_1, b_2, \ldots, b_K}{c_1, c_2, \ldots, c_L}\right] = x^{-q}\left[x \frac{0, b_1, b_2, \ldots, b_K}{-q, c_1, c_2, \ldots, c_L}\right],$$

$$q = 0, \pm 1, \pm 2, \ldots.$$

Of perhaps greater interest is the similarly derived formula

$$(2.10.3) \quad \frac{d^q}{dx^q}\left\{x^p\left[x \frac{b_1, b_2, \ldots, b_K}{c_1, c_2, \ldots, c_L}\right]\right\}$$

$$= x^{p-q}\left[x \frac{p, b_1, b_2, \ldots, b_K}{p-q, c_1, c_2, \ldots, c_L}\right], q = 0, \pm 1, \ldots,$$

expressing the operation of differintegration of the product of the power x^p with a generalized hypergeometric function, provided p exceeds -1. These operations will normally convert a $\frac{K}{L}$ hypergeometric function to $\frac{K+1}{L+1}$, increasing its complexity. However, if p and q are suitably chosen, cancellation with existing parameters will be possible, leaving the complexity of the hypergeometric function unchanged, or even reducing it to $\frac{K-1}{L-1}$. This device, generalized to embrace differintegration to arbitrary order, will prove invaluable in Chapter 9.

We conclude this section by demonstrating that a generalized hypergeometric function of argument $x^{1/n}$ (n being a positive integer) and complexity $\frac{K}{L}$ may be equated to the sum of n hypergeometrics of argument $n^{n[K-L]}x$ and complexity $\frac{nK}{nL}$. The lengthy proof, which makes use of the Gauss multiplication formula [equation (1.3.10)] and of the abbreviations $B = \sum b_k$, $C = \sum c_l$, and $2D = K - L$ will be found on p. 44. This rather forbidding formula condenses remarkably for small values of n, K, and L. Thus we find, for example,

$$\left[\sqrt{x}\,\frac{}{0}\right] = \sqrt{\pi}\left[\tfrac{1}{4}x\,\frac{}{0, -\tfrac{1}{2}}\right] + \frac{\sqrt{\pi x}}{2}\left[\tfrac{1}{4}x\,\frac{}{\tfrac{1}{2}, 0}\right].$$

$$(2.10.4) \quad \left[x^{1/n} \begin{array}{c} b_1, b_2, \ldots, b_K \\ c_1, c_2, \ldots, c_L \end{array} \right]$$

$$= \sum_{m=0}^{n-1} \sum_{j=0}^{\infty} x^{j+[m/n]} \frac{\prod_{k=1}^{K} \Gamma(nj+1+m+b_k)}{\prod_{l=1}^{L} \Gamma(nj+1+m+c_l)}$$

$$= \sum_{m=0}^{n-1} n^{m/n} \sum_{j=0}^{\infty} x^j \frac{\prod_{k=1}^{K} [\sqrt{2\pi}]^{1-n} [n]^{nj+\frac{1}{2}+m+b_k} \prod_{i=0}^{n-1} \Gamma\left(j+1+\frac{m+b_k-i}{n}\right)}{\prod_{l=1}^{L} [\sqrt{2\pi}]^{1-n} [n]^{nj+\frac{1}{2}+m+c_l} \prod_{i=0}^{n-1} \Gamma\left(j+1+\frac{m+c_l-i}{n}\right)}$$

$$= [2\pi]^{D-nD} n^{B-C+D} \sum_{m=0}^{n-1} n^{2Dm} x^{m/n} \sum_{j=0}^{\infty} n^{2nDj} x^j \frac{\prod_{k=1}^{K} \prod_{i=0}^{n-1} \Gamma\left(j+1+\frac{m+b_k-i}{n}\right)}{\prod_{l=1}^{L} \prod_{i=0}^{n-1} \Gamma\left(j+1+\frac{m+c_l-i}{n}\right)}$$

$$= [2\pi]^{D-nD} n^{B-C+D} \sum_{m=0}^{n-1} n^{2Dm} x^{m/n}$$

$$\times \left[n^{2nD} x \begin{array}{c} \dfrac{m+b_1}{n}, \dfrac{m+b_1-1}{n}, \ldots, \dfrac{m+b_1-n+1}{n} \dfrac{m+b_2}{n}, \ldots, \dfrac{m+b_K-n+1}{n} \\ \dfrac{m+c_1}{n}, \dfrac{m+c_1-1}{n}, \ldots, \dfrac{m+c_1-n+1}{n} \dfrac{m+c_2}{n}, \ldots, \dfrac{m+c_L-n+1}{n} \end{array} \right].$$

CHAPTER 3

FRACTIONAL DERIVATIVES AND INTEGRALS: DEFINITIONS AND EQUIVALENCES

In this chapter we compare several rival definitions of the differintegral of a function f to arbitrary order q. All definitions are, of course, required to yield multiple derivatives and integrals when the order is a positive or negative integer.

Perhaps the least ambiguous symbolism for the value at x of the differintegral to order q of a function f defined on the interval $a \leq y \leq x$ would be

$$\frac{d^q f(y)}{[d(y-a)]^q}(x).$$

We shall eventually relate this differintegral to an ordinary integral in which y is a "dummy" variable of integration, and a and x are limits of integration. In line with conventions adopted in Section 2.1, our normal abbreviations for the qth differintegral of a function f will be

$$\frac{d^q f}{[d(x-a)]^q}, \quad \text{and} \quad \frac{d^q f}{[d(x-a)]^q}(x_0) \quad \text{or} \quad \left[\frac{d^q f}{[d(x-a)]^q}\right]_{x=x_0},$$

it being understood that f and $d^q f/[d(x-a)]^q$ are functions of the independent variable x when the x is omitted. Since, as we shall see ultimately, differintegrals must include integral transforms as well as ordinary derivatives and integrals, for which a variety of conflicting notations abound in the literature, the selection of a single satisfactory replacement for all of these is as much a matter of taste as logic. We have attempted to combine both elements in our choice.

3.1 DIFFERINTEGRABLE FUNCTIONS

It is time now to delineate the class of functions to which we shall apply differintegral operators. We shall, for the most part, work within the framework of classically defined functions rather than distributions (sometimes also called symbolic functions, or generalized functions). For such classically defined functions, we take our clue from the integral calculus and require that our candidate functions be defined on the closed interval $a \leq y \leq x$, that they be bounded everywhere in the half-open interval $a < y \leq x$, and be "better behaved" at the lower limit a than is $[y - a]^{-1}$.

We define the class of "differintegrable series" to be all finite sums of functions, each of which may be represented

$$(3.1.1) \quad f(y) = [y - a]^p \sum_{j=0}^{\infty} a_j [y - a]^{j/n}, \quad a_0 \neq 0, \quad p > -1,$$

as the product of a power of $[y - a]$ and an analytic function of $[y - a]^{1/n}$, n a positive integer.[1] Notice that p has been chosen to ensure that the leading coefficient is nonzero in equation (3.1.1). Such differintegrable series f then satisfy

$$\lim_{y \to a} \{[y - a] f(y)\} = 0$$

(this is what we mean by the phrase "better behaved" at a than $[y - a]^{-1}$) and, in addition to bounded examples, include such functions as $f(y) = [y - a]^{-\frac{1}{2}}$ and $f(y) = \sin(\sqrt{y - a})/[y - a]^{\frac{3}{4}}$. Indeed, most of the special functions of mathematical physics are differintegrable series according to the definition we have given. An important consequence of the representation (3.1.1) is that f may be further decomposed as a finite sum

$$f(y) = [y - a]^p \sum_{j_1=0}^{\infty} a_{j_1}[y - a]^{j_1} + [y - a]^{[np+1]/n} \sum_{j_2=0}^{\infty} a_{j_2}[y - a]^{j_2}$$

$$+ \cdots + [y - a]^{[np+n-1]/n} \sum_{j_n=0}^{\infty} a_{j_n}[y - a]^{j_n}$$

of n differintegrable "units" f_U, each of which is a product of a power (greater than -1) of $y - a$ and a function analytic in $y - a$. The desirability

---

[1] It is likely that arbitrary powers of $[y - a]$ could be treated in the infinite series factor of f. At a minimum, inclusion of such functions would complicate many proofs in Chapter 5 without introducing any new ideas. We therefore restrict our attention to the simpler, more tractable, series.

of this property will become more apparent in Chapter 5 (see, for example, Sections 5.2 and 5.7). Another feature motivating the selected form for differintegrable series is that this form is reproduced upon differintegration to any order (although the restriction $p > -1$ may be violated in the differintegrated series). Notice the analogy with analytic series whose form is reproduced upon differentiation or integration to integer order.

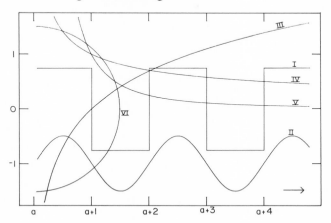

FIG. 3.1.1. Some functions that can be differintegrated: I, a square wave function; II, the sine function $\frac{1}{2}\sin(\pi[x-a])-1$; III, the logarithmic function $\ln(x-a)$; and IV, the inverse power $[x-a]^{-\frac{1}{2}}$ and one that cannot; V, the inverse power $[x-a]^{-2}$. The function VI $\sqrt{\frac{9}{4}-[x-a]^2}$, if treated as single valued, can be differintegrated, though the differintegral will be real only if $a < x \leq \frac{3}{2}+a$.

Many functions not expansible as differintegrable series are, nevertheless, differintegrable. Good examples of such functions are the logarithm and Heaviside's unit function (see Sections 6.7 and 6.8); an even simpler example is the function $f \equiv 0$ (see Section 4.2). When we use the phrase "differintegrable function" we mean to include all of the above and, in fact, any function whose differintegrals can be determined. Figure 3.1.1 shows examples of some functions that can be differintegrated and some that cannot.

3.2 FUNDAMENTAL DEFINITIONS

The first definition we offer is the one we regard as the most fundamental in that it involves the fewest restrictions on the functions to which it applies and avoids explicit use of the notions of ordinary derivative and integral. This definition, which directly extends and unifies notions of difference

48   **3 FRACTIONAL DERIVATIVES AND INTEGRALS**

quotients and Riemann sums, was first given by Grünwald (1867) and later extended by Post (1930).

Referring back to Section 2.2 and the discussion leading up to equation (2.2.3), we shall define the differintegral of order q by the formula

$$(3.2.1) \quad \frac{d^q f}{[d(x-a)]^q} = \lim_{N\to\infty} \left\{ \frac{\left[\frac{x-a}{N}\right]^{-q}}{\Gamma(-q)} \sum_{j=0}^{N-1} \frac{\Gamma(j-q)}{\Gamma(j+1)} f\left(x - j\left[\frac{x-a}{N}\right]\right) \right\},$$

where q is arbitrary.[2] Notice that definition (3.2.1) involves only evaluations of the function itself; no *explicit* use is made of derivatives or integrals of f.

We should like to establish that, based on the definition (3.2.1),

$$(3.2.2) \quad \frac{d^n}{dx^n} \frac{d^q f}{[d(x-a)]^q} = \frac{d^{n+q} f}{[d(x-a)]^{n+q}}$$

for all positive integers n and all q. One might think of this property as a limited composition law, that is, a rule for composing orders of the generalized differintegral. A complete discussion of the composition law will be presented in Section 5.7.

To establish (3.2.2), let $\delta_N x = [x-a]/N$, as in Section 2.2, and notice that

$$\frac{d^q f}{[d(x-a)]^q} = \lim_{N\to\infty} \left\{ \frac{[\delta_N x]^{-q}}{\Gamma(-q)} \sum_{j=0}^{N-1} \frac{\Gamma(j-q)}{\Gamma(j+1)} f(x - j\,\delta_N x) \right\}.$$

Upon subdividing the interval $a \leq y \leq x - \delta_N x$ into only $N-1$ equally spaced subintervals, we see that

$$\frac{d^q f}{[d(x-a)]^q}(x - \delta_N x) = \lim_{N\to\infty} \left\{ \frac{[\delta_N x]^{-q}}{\Gamma(-q)} \sum_{j=0}^{N-2} \frac{\Gamma(j-q)}{\Gamma(j+1)} f(x - \delta_N x - j\,\delta_N x) \right\}$$

$$= \lim_{N\to\infty} \left\{ \frac{[\delta_N x]^{-q}}{\Gamma(-q)} \sum_{j=1}^{N-1} \frac{\Gamma(j-q-1)}{\Gamma(j)} f(x - \delta_N x) \right\}.$$

On differentiation (making use of restricted limits as explained in Section 2.2 to define d/dx), one gets

$$\frac{d}{dx} \frac{d^q f}{[d(x-a)]^q}$$

$$\equiv \lim_{N\to\infty} \left\{ [\delta_N x]^{-1} \left[\frac{d^q f}{[d(x-a)]^q}(x) - \frac{d^q f}{[d(x-a)]^q}(x - \delta_N x) \right] \right\}$$

$$= \lim_{N\to\infty} \left\{ \frac{[\delta_N x]^{-q-1}}{\Gamma(-q)} \left[\Gamma(-q) f(x) + \sum_{j=1}^{N-1} \left\{ \frac{\Gamma(j-q)}{\Gamma(j+1)} - \frac{\Gamma(j-q-1)}{\Gamma(j)} \right\} \right] \right\}.$$

[2] Even for q a nonnegative integer [so that $\Gamma(-q)$ is infinite] the ratio $\Gamma(j-q)/\Gamma(-q)$ is finite.

3.2 FUNDAMENTAL DEFINITIONS

But making use of the recurrence properties of gamma functions [equation (1.3.2)],

$$\frac{\Gamma(j-q)}{\Gamma(j+1)} - \frac{\Gamma(j-q-1)}{\Gamma(j)} = \frac{\Gamma(-q)\Gamma(j-q-1)}{\Gamma(-q-1)\Gamma(j+1)}$$

is obtained. Therefore,

$$\frac{d}{dx}\frac{d^q f}{[d(x-a)]^q} = \lim_{N\to\infty}\left\{\frac{[\delta_N x]^{-q-1}}{\Gamma(-q-1)}\left[\sum_{j=0}^{N-1}\frac{\Gamma(j-q-1)}{\Gamma(j+1)}f(x-j\,\delta_N x)\right]\right\}$$
$$= \frac{d^{q+1}f}{[d(x-a)]^{q+1}}.$$

Equation (3.2.2) follows by induction.

The most frequently encountered definition of an integral of fractional order is via an integral transform called the Riemann–Liouville integral (Liouville, 1832a; Riesz, 1949; Riemann, 1953). To motivate this definition, one need only examine Cauchy's formula (2.7.2) and replace $-n$ by q, suggesting the generalization to noninteger q:

(3.2.3) $$\left[\frac{d^q f}{[d(x-a)]^q}\right]_{R-L} = \frac{1}{\Gamma(-q)}\int_a^x [x-y]^{-q-1} f(y)\,dy, \qquad q < 0.$$

In equation (3.2.3) we have used the symbol $[\ ]_{R-L}$ to designate the Riemann–Liouville fractional integral as possibly distinct from our more basic definition (3.2.1). Presently we shall show that the two definitions yield identical results and the symbol $[\ ]_{R-L}$ will be dropped at that point.

Riesz (1949) regarded q as a complex variable; the integral (3.2.3) converges for $\text{Re}(q) < 0$ and defines, for fixed f, an analytic function of q in the left-half q-plane. Although the integral in (3.2.3) diverges when $\text{Re}(q) \geqq 0$, a meaning may be attached to the operator $[\ ]_{R-L}$ by a proper analytic continuation across the line $\text{Re}(q) = 0$ provided the function f is sufficiently differentiable in $a \leqq y \leqq x$. In fact, if f is n times differentiable the formula

(3.2.4) $$\left[\frac{d^q f}{[d(x-a)]^q}\right]_{R-L} = \sum_{k=0}^{n-1}\frac{[x-a]^{-q+k}f^{(k)}(a)}{\Gamma(-q+k+1)} + \left[\frac{d^{q-n}f^{(n)}}{[d(x-a)]^{q-n}}\right]_{R-L}$$

defines an analytic function of q for $\text{Re}(q) < n$ and thus provides a valid analytic continuation [see, for example, Knopp (1945)] of formula (3.2.3) (Riesz, 1949; Duff, 1956). This analytic continuation is very analogous to the way one extends the definition of the gamma function (see Section 1.3) in the complex plane. This is not surprising since the gamma function plays such a fundamental role in defining differintegrals. We shall frequently make use of this analyticity in q to simplify proofs dealing with $d^q f/[d(x-a)]^q$ for general q.

We prefer, however, to take a different, more elementary, approach in attaching a meaning to the operator for $q \geq 0$. Formula (3.2.3) will be retained as the $q < 0$ definition of the differintegral; it is extended to $q \geq 0$ by insisting that equation (3.2.2) be satisfied by the Riemann–Liouville integral. That is, we shall require that

$$(3.2.5) \qquad \left[\frac{d^q f}{[d(x-a)]^q}\right]_{R-L} \equiv \frac{d^n}{dx^n}\left[\frac{d^{q-n} f}{[d(x-a)]^{q-n}}\right]_{R-L},$$

where d^n/dx^n effects ordinary n-fold differentiation and n is an integer chosen so large that $q - n < 0$. Together with equation (3.2.5), definition (3.2.3) then defines the operator

$$\left[\frac{d^q}{[d(x-a)]^q}\right]_{R-L}$$

for all q.

We see that if we choose q to equal the negative integer $-n$ in equation (3.2.3), we obtain

$$\left[\frac{d^{-n} f}{[d(x-a)]^{-n}}\right]_{R-L} = \frac{1}{\Gamma(n)}\int_a^x [x-y]^{n-1} f(y)\,dy.$$

Comparison with equation (2.7.2) reveals that the Riemann–Liouville definition correctly generates an n-fold integral of f. It is also evident on choosing $n = 1$ and $q = 0$ in equation (3.2.5) that

$$\left[\frac{d^0 f}{[d(x-a)]^0}\right]_{R-L} = \frac{d}{dx}\left[\frac{d^{-1} f}{[d(x-a)]^{-1}}\right]_{R-L} = f.$$

Moreover, by selecting $n = q$, we establish that

$$\left[\frac{d^n f}{[d(x-a)]^n}\right]_{R-L} = \frac{d^n f}{dx^n}$$

is the ordinary nth derivative when n is a nonnegative integer.

Courant and Hilbert (1962) defined a semiderivative by

$$(3.2.6) \qquad \frac{d^{\frac{1}{2}} f}{[d(x-a)]^{\frac{1}{2}}} \equiv \frac{1}{\sqrt{\pi}}\frac{d}{dx}\int_a^x \frac{f(y)\,dy}{\sqrt{x-y}} = \frac{d}{dx}\frac{d^{-\frac{1}{2}} f}{[d(x-a)]^{-\frac{1}{2}}},$$

which is the same as our extended Riemann–Liouville definition, as is seen from equation (3.2.5). A careful integration by parts shows that

$$(3.2.7) \qquad \left[\frac{d^{\frac{1}{2}} f}{[d(x-a)]^{\frac{1}{2}}}\right]_{R-L} = \frac{1}{\sqrt{\pi}}\frac{d}{dx}\int_a^x \frac{f(y)\,dy}{\sqrt{x-y}}$$

$$= \frac{1}{\sqrt{\pi}}\frac{f(a)}{\sqrt{x-a}} + \frac{1}{\sqrt{\pi}}\int_a^x \frac{f^{(1)}(y)\,dy}{\sqrt{x-y}}$$

$$= \frac{f(a)}{\Gamma(\frac{1}{2})\sqrt{x-a}} + \left[\frac{d^{-\frac{1}{2}} f^{(1)}}{[d(x-a)]^{-\frac{1}{2}}}\right]_{R-L},$$

which agrees with equation (3.2.4), as it must, for $q = +\frac{1}{2}$ and $n = 1$. We notice, however, that a semiderivative defined in this way requires explicit knowledge of the first derivative $f^{(1)}$. On the other hand, setting $q = \frac{1}{2}$ in our fundamental definition (3.2.1) avoids explicit use of the first derivative, as does equation (3.2.6).

3.3 IDENTITY OF DEFINITIONS

It is now pertinent to ask whether the Riemann–Liouville definition, based on equation (3.2.3) for negative q and its extension to $q \geq 0$ by means of equation (3.2.5) is equivalent to the definition (3.2.1). That is, do the operators so defined coincide for all functions f? We shall establish that this is, indeed, the case. First we prove the identity for a subset of q values[3] and then make use of the property (3.2.2) to extend the identity to all orders q.

Thus we choose f to be an arbitrary, but fixed, function on the interval $a \leq y \leq x$. As before, let $\delta_N x = [x - a]/N$. Then the difference

$$\Delta \equiv \frac{d^q f}{[d(x-a)]^q}(x) - \left[\frac{d^q f}{[d(x-a)]^q}(x)\right]_{R-L}$$

$$= \lim_{N \to \infty} \left\{ \frac{[\delta_N x]^{-q}}{\Gamma(-q)} \sum_{j=0}^{N-1} \frac{\Gamma(j-q)}{\Gamma(j+1)} f(x - j\,\delta_N x) \right\} - \int_0^{x-a} \frac{f(x-u)\,du}{\Gamma(-q) u^{1+q}}$$

$$= \lim_{N \to \infty} \left\{ \frac{[\delta_N x]^{-q}}{\Gamma(-q)} \sum_{j=0}^{N-1} \frac{\Gamma(j-q)}{\Gamma(j+1)} f(x - j\,\delta_N x) \right\} - \lim_{N \to \infty} \left\{ \sum_{j=0}^{N-1} \frac{f(x - j\,\delta_N x)\,\delta_N x}{\Gamma(-q)[j\,\delta_N x]^{1+q}} \right\}$$

$$= \lim_{N \to \infty} \left\{ \frac{[\delta_N x]^{-q}}{\Gamma(-q)} \sum_{j=0}^{N-1} f(x - j\,\delta_N x) \left[\frac{\Gamma(j-q)}{\Gamma(j+1)} - j^{-1-q} \right] \right\}$$

$$= \frac{[x-a]^{-q}}{\Gamma(-q)} \lim_{N \to \infty} \left\{ \sum_{j=0}^{N-1} f\left(\frac{Nx - jx + ja}{N}\right) N^q \left[\frac{\Gamma(j-q)}{\Gamma(j+1)} - j^{-1-q} \right] \right\}.$$

The N terms within the summation will be treated as two groups: $0 \leq j \leq J - 1$ and $J \leq j \leq N - 1$, where J is independent of N and large enough to validate the asymptotic expansion (1.3.13) for summands in the second group. Thus

$$\Delta = \frac{[x-a]^{-q}}{\Gamma(-q)} \lim_{N \to \infty} \left\{ \sum_{j=0}^{J-1} f\left(\frac{Nx - jx + ja}{N}\right) N^q \left[\frac{\Gamma(j-q)}{\Gamma(j+1)} - j^{-1-q} \right] \right\}$$

$$+ \frac{[x-a]^{-q}}{\Gamma(-q)} \lim_{N \to \infty} \left\{ \frac{1}{N} \sum_{j=J}^{N-1} f\left(\frac{Nx - jx + ja}{N}\right) \left[\frac{j}{N}\right]^{-2-q} \left[\frac{q[q+1]}{2N} + \frac{O(j^{-1})}{N} \right] \right\}.$$

---

[3] For integer q of either sign this identity is intrinsic to both the Grünwald and the Riemann–Liouville definitions.

52 3 FRACTIONAL DERIVATIVES AND INTEGRALS

Now, for $q < -1$, the J bracketed terms within the first summation are bounded. Hence, if $f([Nx - jx + ja]/N)$ is also bounded for j within the first group, the presence of the N^q factor ensures that the first sum vanishes in the $N \to \infty$ limit. Examining the three factors within the second summation, we note that $[j/N]^{-2-q}$ is invariably less than unity if $q \leq -2$ and that the third factor tends to zero when $N \to \infty$. Hence, if $f([Nx - jx + ja]/N)$ is bounded for j within the second group, each term in the second summation vanishes as $1/N$ when $N \to \infty$. As there are fewer than N summands in the second group, the presence of the pre-summation $1/N$ factor ensures that the second limit is zero.

The above demonstrates that if f is bounded on $a < y \leq x$ and if $q \leq -2$, then

$$\frac{d^q f}{[d(x-a)]^q}(x) - \left[\frac{d^q f}{[d(x-a)]^q}(x)\right]_{R-L} \equiv \Delta = 0, \quad (3.3.1)$$

so that the two definitions when applied to functions so bounded[4] are indeed identical for $q \leq -2$. This fact, coupled with the property (3.2.2) and the requirement (3.2.5), shows that the two definitions are identical for any q. Indeed, for arbitrary q we know that for any positive integer n

$$\frac{d^q f}{[d(x-a)]^q} = \frac{d^n}{dx^n}\left\{\frac{d^{q-n} f}{[d(x-a)]^{q-n}}\right\}$$

and

$$\left[\frac{d^q f}{[d(x-a)]^q}\right]_{R-L} = \frac{d^n}{dx^n}\left[\frac{d^{q-n} f}{[d(x-a)]^{q-n}}\right]_{R-L}.$$

One need only choose n sufficiently large that $q - n \leq -2$ and make use of (3.3.1) to complete the proof.

3.4 OTHER GENERAL DEFINITIONS

We have presented in Section 3.2 the two definitions of generalized differintegration which we favor. Other starting points for generalizing ordinary derivatives and integrals have been used in the literature, however. In the present section some of these are discussed briefly. Our treatment is

---

[4] The astute reader may have noticed that, while our argument was made only for functions bounded on $a < y \leq x$, the proof may be sharpened to admit all differintegrable series (see Section 3.1). We have omitted the rather tedious proof.

3.4 OTHER GENERAL DEFINITIONS

cursory rather than exhaustive since we shall have little occasion throughout the rest of this book to refer to any definitions other than those of Section 3.2.

Riemann (1953) considered power series with noninteger exponents to be extensions of Taylor's series and built up a generalized derivative for such functions by use of the formula

(3.4.1) $$\frac{d^q x^p}{dx^q} = \frac{\Gamma(p+1)}{\Gamma(p-q+1)} x^{p-q},$$

this being an obvious generalization of the formula

$$\frac{d^n x^p}{dx^n} = p[p-1][p-2] \cdots [p-n+1] x^{p-n} = \frac{\Gamma(p+1)}{\Gamma(p-n+1)} x^{p-n}$$

for n a nonnegative integer. This is similar to the approach taken by Scott Blair (1947), Heaviside (1920), and others. Liouville (1832a) defined a generalized derivative for functions expansible as a series of exponentials, $f = \sum c_j \exp(b_j x)$, by

$$\frac{d^q f}{dx^q} \equiv \sum_{j=0}^{\infty} c_j b_j^q \exp(b_j x),$$

a definition that leads to a different operator than those of Section 3.2. Krug (1890), in fact, showed that this definition corresponds to a lower limit $a = -\infty$ in the Riemann–Liouville integral (3.2.3). Civin (1941) used a similar idea for integrals rather than sums to define

$$\frac{d^q f}{dx^q} \equiv \int_{-\pi}^{\pi} i^q t^q \exp(ixt) \, ds(t),$$

where f is given by

$$f \equiv \int_{-\pi}^{\pi} \exp(ixt) \, ds(t).$$

Weyl (1917), working with periodic functions, defined the qth differintegral of f via

$$\frac{d^q f}{dx^q} \equiv \int_{-\infty}^{x} \frac{f(y) \, dy}{[x-y]^{1+q}},$$

this integral being often called the Weyl integral. More recently, Gaer (1968) and Gaer and Rubel (1971) have given a definition closely related to the Weyl integral. Based on their definition, Gaer and Rubel developed a formula for the fractional derivative of a product which does not reduce to Leibniz's formula in the classical case. We shall have more to say about extensions of Leibniz's formula in Section 5.5.

3 FRACTIONAL DERIVATIVES AND INTEGRALS

A different avenue for motivating the definition of a differintegral stems from consideration of Cauchy's integral formula:

$$(3.4.2) \qquad \frac{d^n f(z)}{dz^n} = \frac{n!}{2\pi i} \oint_C \frac{f(\zeta)\, d\zeta}{[\zeta - z]^{n+1}},$$

where C describes a closed contour surrounding the point z and enclosing a region of analyticity of f. When the positive integer n is replaced by a noninteger q, then $[\zeta - z]^{-q-1}$ no longer has a pole at $\zeta = z$ but a branch point. One is no longer free to deform the contour C surrounding z at will, since the integral will depend on the location of the point at which C crosses the branch line for $[\zeta - z]^{-q-1}$. This point is chosen to be 0 and the branch line to be the straight line joining 0 and z and continuing indefinitely in the quadrant $\mathrm{Re}(\zeta) \leq 0$, $\mathrm{Im}(\zeta) \leq 0$. Then one simply defines, for q not a negative integer,

$$(3.4.3) \qquad \frac{d^q f}{dz^q} = \frac{\Gamma(q+1)}{2\pi i} \oint_C \frac{f(\zeta)\, d\zeta}{[\zeta - z]^{q+1}},$$

where the contour C begins and ends at $\zeta = 0$ enclosing z once in the positive sense. To uniquely specify the denominator of the integrand, one defines

$$[\zeta - z]^{q+1} = \exp([q+1]\ln(\zeta - z)),$$

where $\ln(\zeta - z)$ is real when $\zeta - z > 0$. We can relate the definition (3.4.3) to that of Riemann–Liouville by first deforming the contour C into a contour C' lying on both sides of the branch line (see Fig. 3.4.1). One finds that[5]

$$\frac{\Gamma(q+1)}{2\pi i} \oint_{C'} \frac{f(\zeta)\, d\zeta}{[\zeta - z]^{q+1}} = \frac{\Gamma(q+1)}{2\pi i}\left[1 - \exp(-2\pi i[q+1])\right]\int_0^z \frac{f(\zeta)\, d\zeta}{[\zeta - z]^{q+1}}$$

$$= \frac{1}{\Gamma(-q)} \int_0^z \frac{f(\zeta)\, d\zeta}{[z - \zeta]^{q+1}},$$

which is the Riemann–Liouville definition (3.2.3) with $a = 0$. The definition (3.4.3) is attributed by Osler (1970a) to Nekrassov (1888).

Erdélyi (1964) defined a qth-order differintegral of a function $f(z)$ with respect to the function z^n by

$$\frac{d^q f}{[d(z^n - a^n)]^q} \equiv \frac{1}{\Gamma(-q)} \int_a^z \frac{f(\zeta)n\zeta^{n-1}\, d\zeta}{[z^n - \zeta^n]^{1+q}}.$$

---

[5] In establishing the second equality we have used $[-1]^{-q-1} = \exp(i\pi[q+1])$ and the reflection formula $\Gamma(x)\Gamma(1-x) = \pi\csc(\pi x) = 2\pi i/[\exp(\pi i x) - \exp(-\pi i x)]$ for the gamma function.

3.4 OTHER GENERAL DEFINITIONS

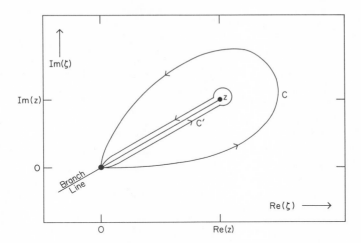

FIG. 3.4.1 The Cauchy contour C is deformed into C' for the purpose of implementing definition (3.4.3).

Osler (1970a) has extended Erdélyi's work by defining a differintegral of a function $f(z)$ with respect to an arbitrary function $g(z)$ by considering the Riemann–Liouville integral

$$\frac{d^q f}{[d(g(z) - g(a))]^q} \equiv \frac{1}{\Gamma(-q)} \int_a^z \frac{f(\zeta) g^{(1)}(\zeta)\, d\zeta}{[g(z) - g(\zeta)]^{q+1}},$$

where a is chosen to give $g(a) = 0$, that is $a = g^{-1}(0)$. Upon setting $g(z) = z - a$, one obtains the Riemann–Liouville integral once again. Making the change $u \equiv g(\zeta)/g(z)$ gives

$$\frac{d^q f}{[d(g(z) - g(a))]^q} = \frac{g(z)^{-q}}{\Gamma(-q)} \int_0^1 \frac{f(g^{-1}(g(z)u))}{[1 - u]^{q+1}}\, du.$$

Certain choices of g have been shown by Erdélyi and by Osler to lead to a number of formulas of interest in classical analysis.

We close this section with a short discussion of the pros and cons of the various definitions of generalized differintegrals. The fundamental definition (3.2.1) based on difference quotients and Riemann sums has the great merit of general applicability but is awkward to use as a working definition except for fairly simple functions. We favor it as a basic definition because the functions to which it may be applied are not limited in any intrinsic way, nor does it explicitly make use of the classical notions of derivative or integral. It may also have merit as a tool for generating numerical approximations to the differintegral of an arbitrary function.

The Riemann–Liouville definition (3.2.3) is convenient to implement but requires $q < 0$ for the integral to converge. Application of (3.2.5) to circumvent this restriction may also lead to difficulties with implementation since one must compute an n-fold derivative of an integral. Nevertheless, because of its convenient formulation in terms of a single integral it enjoys great popularity as a working definition of a differintegral.

The definition (3.4.3) based on Cauchy's integral formula is, as we have seen, closely related to the Riemann–Liouville definition. However, it applies only to locally analytic functions and is inapplicable when q is a negative integer. A desirable feature of the definition (3.4.3) is the flexibility provided through the choice of the contour C. Making use of this flexibility, Osler (1972a) has extended several notions valid for real-valued functions "into the complex plane." Thus he has used (3.4.3) to develop an integral analog of Taylor's series and also to develop an integral analog of Lagrange's expansion, which he shows extends some features of Fourier analysis into the complex plane. Indeed, an appropriate choice of contour C reveals that, apart from a constant factor, $z^q d^q f/dz^q$ is essentially the Fourier transform in q-space of the function f. Many useful results of Fourier analysis may be developed from this starting point by making use of the general properties of differintegral operators.

Before leaving this section, let us return briefly to the Grünwald definition. Recall that the motivation for definition (3.2.1),

$$\frac{d^q f}{[d(x-a)]^q} = \lim_{N \to \infty} \left\{ \left[\frac{x-a}{N}\right]^{-q} \frac{1}{\Gamma(-q)} \sum_{j=0}^{N-1} \frac{\Gamma(j-q)}{\Gamma(j+1)} f\left(x - j\left[\frac{x-a}{N}\right]\right) \right\},$$

is that it correctly reproduces standard classical definitions when q is an integer of either sign. Thus when $q = -1$ or $+1$, to take the simplest examples, the Grünwald definition reduces respectively to a Riemann sum limit,

(3.4.4) $$\frac{d^{-1} f}{[d(x-a)]^{-1}} = \lim_{N \to \infty} \left\{ \frac{x-a}{N} \sum_{j=0}^{N-1} f\left(x - j\left[\frac{x-a}{N}\right]\right) \right\},$$

or a backward difference quotient limit,

(3.4.5) $$\frac{d^1 f}{[d(x-a)]^1} = \lim_{N \to \infty} \left\{ \left[\frac{x-a}{N}\right]^{-1} \sum_{j=0}^{1} [-1]^j f\left(x - j\left[\frac{x-a}{N}\right]\right) \right\}.$$

Notice, however, that the Riemann sum would converge more rapidly to the integral as $N \to \infty$ were formula (3.4.4) replaced by

$$\frac{d^{-1} f}{[d(x-a)]^{-1}} = \lim_{N \to \infty} \left\{ \frac{x-a}{N} \sum_{j=0}^{N-1} f\left(x - [j + \tfrac{1}{2}]\left[\frac{x-a}{N}\right]\right) \right\}.$$

3.5 OTHER FORMULAS APPLICABLE TO ANALYTIC FUNCTIONS

Similarly, the difference quotient would converge more rapidly to the true derivative were (3.4.5) replaced by the central difference formulation

$$\frac{d^1 f}{[d(x-a)]^1} = \lim_{N\to\infty} \left\{ \left[\frac{x-a}{N}\right]^{-1} \sum_{j=0}^{1} [-1]^j f\left(x - [j-\tfrac{1}{2}]\left[\frac{x-a}{N}\right]\right) \right\}.$$

Similar modifications improve the Grünwald definition for other integer q values and suggest a generalization to

$$(3.4.6) \quad \frac{d^q f}{[d(x-a)]^q} = \lim_{N\to\infty} \left\{ \frac{\left[\dfrac{x-a}{N}\right]^{-q}}{\Gamma(-q)} \sum_{j=0}^{N-1} \frac{\Gamma(j-q)}{\Gamma(j+1)} f\left(x - [j-\tfrac{1}{2}q]\left[\frac{x-a}{N}\right]\right) \right\}.$$

The improved convergence of this modified Grünwald formula permits the design of a more efficient differintegration algorithm, as we shall see in Section 8.2.

3.5 OTHER FORMULAS APPLICABLE TO ANALYTIC FUNCTIONS

The purpose of this section is to discuss alternate representations for $d^q/[d(x-a)]^q$ for real analytic functions, i.e., functions ϕ that have convergent power series expansions in the interval $a \leq y \leq x$ of interest. Such representations offer computational variety when it comes to the evaluation of qth order differintegrals for specific choices of ϕ.

We restrict attention initially to $q < 0$, since this permits us to work with the Riemann–Liouville definition (3.2.3). Thus

$$(3.5.1) \quad \frac{d^q \phi}{[d(x-a)]^q} = \frac{1}{\Gamma(-q)} \int_a^x \frac{\phi(y)\,dy}{[x-y]^{q+1}} = \frac{1}{\Gamma(-q)} \int_0^{x-a} \frac{\phi(x-v)\,dv}{v^{q+1}}$$

with $v \equiv x - y$. Upon Taylor expansion of $\phi(x-v)$ about x one has

$$(3.5.2) \quad \phi(x-v) = \phi - v\phi^{(1)} + \frac{v^2}{2!}\phi^{(2)} - \cdots = \sum_{k=0}^{\infty} \frac{[-1]^k v^k \phi^{(k)}}{k!}.$$

The representation (3.5.2) involves no remainder since we have assumed ϕ to have a convergent power series expansion and since such an expansion is unique. When this expansion is inserted into (3.5.1) and term-by-term integration performed, the result is

$$(3.5.3) \quad \frac{d^q \phi}{[d(x-a)]^q} = \sum_{k=0}^{\infty} \frac{[-1]^k [x-a]^{k-q} \phi^{(k)}}{\Gamma(-q)[k-q]k!}.$$

3 FRACTIONAL DERIVATIVES AND INTEGRALS

An analyticity (in q) argument[6] may be used to establish the formula (3.5.3) for all q even though it was derived on the basis of the assumption $q < 0$.

Another proof of equation (3.5.3) that starts with definition (3.2.1) proceeds as follows, making use of the equation (1.3.32):

$$\frac{d^q \phi}{[d(x-a)]^q}$$

$$= [x-a]^{-q} \lim_{N \to \infty} \left\{ N^q \sum_{j=0}^{N-1} \frac{\Gamma(j-q)}{\Gamma(-q)\Gamma(j+1)} \phi\left(x + j\left[\frac{a-x}{N}\right]\right) \right\}$$

$$= [x-a]^{-q} \lim_{N \to \infty} \left\{ N^q \sum_{j=0}^{N-1} \binom{j-q-1}{j} \sum_{m=0}^{\infty} \binom{j}{m} G_m\left(\phi, x, \frac{a-x}{N}\right) \right\}$$

$$= [x-a]^{-q} \lim_{N \to \infty} \left\{ N^q \sum_{m=0}^{\infty} G_m\left(\phi, x, \frac{a-x}{N}\right) \sum_{j=0}^{N-1} \binom{j-q-1}{j}\binom{j}{m} \right\}$$

$$= [x-a]^{-q} \lim_{N \to \infty} \left\{ N^q \sum_{m=0}^{\infty} \binom{m-q-1}{m}\binom{N-q-1}{N-m-1} G_m\left(\phi, x, \frac{a-x}{N}\right) \right\},$$

where the final step is a consequence of formula (1.3.22). Now we use the expression for $G_m(\phi, x, y)$ given in (1.3.33) to get

$$\frac{d^q \phi}{[d(x-a)]^q}$$

$$= [x-a]^{-q} \lim_{N \to \infty} \left\{ N^q \sum_{m=0}^{\infty} m! \binom{m-q-1}{m}\binom{N-q-1}{N-m-1} \right.$$

$$\left. \times \sum_{k=m}^{\infty} \left[\frac{a-x}{N}\right]^k S_k^{[m]} \frac{\phi^{(k)}(x)}{k!} \right\}$$

$$= \sum_{m=0}^{\infty} \frac{\Gamma(m-q)}{\Gamma(-q)} \sum_{k=m}^{\infty} [-]^k \frac{[x-a]^{k-q}}{k!} S_k^{[m]} \phi^{(k)}(x) \lim_{N \to \infty} \left\{ N^{q-k}\binom{N-q-1}{N-m-1} \right\}.$$

The summation on k involves only one nonzero term by virtue of

$$\lim_{N \to \infty} \left\{ N^{q-k}\binom{N-q-1}{N-m-1} \right\} = \begin{cases} \dfrac{1}{\Gamma(m-q+1)}, & k=m, \\ 0, & k>m, \end{cases}$$

which follows from expression (1.3.15). This results in

$$(3.5.4) \quad \frac{d^q \phi}{[d(x-a)]^q} = \sum_{m=0}^{\infty} \frac{\Gamma(m-q)}{\Gamma(-q)} \frac{[-]^m [x-a]^{m-q}}{m!} S_m^{[m]} \frac{\phi^{(m)}(x)}{\Gamma(m-q+1)},$$

[6] See p. 49.

which, by virtue of equation (1.3.29) and the recurrence property (1.3.2) of the gamma function, is seen to identify with equation (3.5.3).

We have presented this alternative proof of formula (3.5.3) to illustrate how it is possible to handle definition (3.2.1). In general, however, as with this example, the algebraic manipulation required to utilize definition (3.2.1) exceeds that required for an approach that employs the Riemann–Liouville definition followed by an analyticity argument. Accordingly, the latter method will constitute our usual approach to the development of formulas for differintegrals.

Formula (3.5.3) is useful inasmuch as it involves only integer-order derivatives of $\phi(x)$. This formula may be written somewhat more concisely, as will be demonstrated in Section 4.1.

3.6 SUMMARY OF DEFINITIONS

In summary, this chapter has shown that

(3.6.1) $$\lim_{N \to \infty} \left\{ \left[\frac{N}{x-a} \right]^q \sum_{j=0}^{N-1} \frac{\Gamma(j-q)}{\Gamma(-q)\Gamma(j+1)} f\left(\frac{Nx - jx + ja}{N} \right) \right\}$$

is a valid expression for a qth order derivative or a $(-q)$th order integral of f whether or not q is a real integer, and that

(3.6.2) $$\frac{1}{\Gamma(-q)} \int_a^x \frac{f(y)\, dy}{[x-y]^{q+1}}$$

is identical with (3.6.1) for $q < 0$. To replace (3.6.2) for $q \geq 0$ one must utilize either

(3.6.3) $$\frac{d^n}{dx^n} \left[\frac{1}{\Gamma(n-q)} \int_a^x \frac{f(y)\, dy}{[x-y]^{q-n+1}} \right], \quad n > q,$$

or

(3.6.4) $$\sum_{k=0}^{n-1} \frac{[x-a]^{k-q} f^{(k)}(a)}{\Gamma(k-q+1)} + \frac{1}{\Gamma(n-q)} \int_a^x \frac{f^{(n)}(y)\, dy}{[x-y]^{q-n+1}}, \quad n > q,$$

both these forms yielding results identical with those given by (3.6.1).

If ϕ is analytic, the expression

(3.6.5) $$\sum_{k=0}^{\infty} \frac{[-]^k [x-a]^{k-q} \phi^{(k)}}{\Gamma(-q)[k-q]k!},$$

developed in Section 3.5, is identical with those above. We shall prove in Section 5.2 that the expression

$$\text{(3.6.6)} \qquad \sum_{k=0}^{\infty} \frac{[x-a]^{k-q}\phi^{(k)}(a)}{\Gamma(k-q+1)}$$

is also a valid representation of $d^q\phi/[d(x-a)]^q$. We notice that (3.6.6) is obtained from (3.6.4) by letting n tend to infinity and dropping the integral remainder term.

It is appropriate to restate that the first four formulas define

$$\frac{d^q f}{[d(x-a)]^q}$$

for all q (real or complex, integer or fractional) and for all differintegrable[7] f. The last two formulas have been established only for analytic functions ϕ. The operation of differintegration is thereby defined between the limits a and x, where $x > a$. Evaluation at the point $x = a$ is specifically excluded.

[7] See Section 3.1 for a discussion concerning the notion of a differintegrable function.

CHAPTER 4

DIFFERINTEGRATION OF SIMPLE FUNCTIONS

It is the purpose of this chapter to calculate the qth-order differintegral of certain simple functions. The formulas we develop will play a major role in all our later work. Because of the identities established in Chapter 3, we are at liberty to use whichever of the definitions summarized in Section 3.6 we find most convenient for a given function. We shall even find it convenient to use, for a given f, different formulas for different q ranges.

The simple functions that are considered in three of the sections of this chapter are examples of the power functions $[x - a]^p$. Thus in Sections 4.1 and 4.3 we treat the $p = 0$ and $p = 1$ instances of this function, while in Section 4.4 the general case is examined. We refer the reader back to Section 2.9 for a review of the information available from the classical calculus.

4.1 THE UNIT FUNCTION

We consider first the differintegral to order q of the function $f \equiv 1$, for which we find it convenient to reserve the special notation [1]. We shall refer to this function as the unit function and trust that no confusion will arise between [1] and the unit *step* function: $f = 0$, $-\infty \leq x < a$; $f = 1$, $x \geq a$ [which we shall term the Heaviside function $H(x - a)$ and shall encounter in Section 6.9].

A straightforward application of (3.2.1) to the function [1] gives

$$\frac{d^q[1]}{[d(x - a)]^q} = \lim_{N \to \infty} \left\{ \left[\frac{N}{x - a}\right]^q \sum_{j=0}^{N-1} \frac{\Gamma(j - q)}{\Gamma(-q)\Gamma(j + 1)} \right\}.$$

4 DIFFERINTEGRATION OF SIMPLE FUNCTIONS

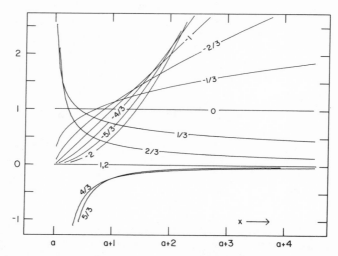

FIG. 4.1.1. Differintegrals of unity for q values in the range -2 to $+2$. Notice that $d^q[1]/[d(x-a)]^q$ is zero for all positive integer q; negative for $1 < q < 2$, $3 < q < 4$, etc.; and otherwise positive.

Application first of equation (1.3.18) and then of (1.3.14) yields

$$(4.1.1) \quad \frac{d^q[1]}{[d(x-a)]^q} = \lim_{N \to \infty} \left\{ \left[\frac{N}{x-a}\right]^q \frac{\Gamma(N-q)}{\Gamma(1-q)\Gamma(N)} \right\} = \frac{[x-a]^{-q}}{\Gamma(1-q)}$$

as our result. Figure 4.1.1 shows some examples of these differintegrals. We note the reduction to an instance of (2.9.3) when q is an integer. Formula (4.1.1) was derived (in a different symbolism and context) in the 1890's by Heaviside (1920).

As an example of the application of the unit function and its differintegrals, consider the combination of formulas (3.6.5) and (4.1.1) into

$$\frac{d^q \phi}{[d(x-a)]^q} = \sum_{k=0}^{\infty} [-]^k \frac{\Gamma(1+k-q)}{\Gamma(-q)[k-q]k!} \frac{d^{q-k}[1]}{[d(x-a)]^{q-k}} \phi^{(k)}$$

valid for any analytic function ϕ. Application of a number of the properties of the gamma function [equations (1.3.2), (1.3.3), and (1.3.16)] then leads to the concise representation

$$(4.1.2) \quad \frac{d^q \phi}{[d(x-a)]^q} = \sum_{k=0}^{\infty} \binom{q}{k} \frac{d^{q-k}[1]}{[d(x-a)]^{q-k}} \frac{d^k \phi}{[d(x-a)]^k},$$

reminiscent of Leibniz's theorem

$$\frac{d^n}{dx^n}[fg] = \sum_{k=0}^{n} \binom{n}{k} \frac{d^{n-k}f}{dx^{n-k}} \frac{d^k g}{dx^k}$$

for differentiation of a product, except that in (4.1.2) we have chosen $f \equiv 1$ but have allowed the order to be unrestricted. The result, as we might expect, is an infinite rather than a finite sum of terms. We have more to say on the subject of generalizations of Leibniz's theorem in Section 5.5.

4.2 THE ZERO FUNCTION

When the definition (3.6.1) is applied to the function defined by $f \equiv C$, C any constant including zero, we see that

$$(4.2.1) \qquad \frac{d^q[C]}{[d(x-a)]^q} = C \frac{d^q[1]}{[d(x-a)]^q} = C \frac{[x-a]^{-q}}{\Gamma(1-q)}.$$

Since $d^q[1]/[d(x-a)]^q$ is never infinite for $x > a$, we conclude by setting $C = 0$ that

$$(4.2.2) \qquad \frac{d^q[0]}{[d(x-a)]^q} = 0 \qquad \text{for all } q.$$

Result (4.2.2) may appear trivial or obvious. As an example of its importance, however, observe that it provides a powerful counterexample to the thesis that if

$$\frac{d^q f}{[d(x-a)]^q} = g, \qquad \text{then} \qquad \frac{d^{-q} g}{[d(x-a)]^{-q}} = f,$$

for, if f gives zero on differentiation to order q, f cannot be restored by q-order integration. Here again, as in Section 2.3, we encounter the so-called composition rule, this time for noninteger orders. This subject will be more fully explored in Section 5.7.

4.3 THE FUNCTION $x-a$

For the function $f(x) = x - a$, definition (3.6.1) gives

$$\frac{d^q[x-a]}{[d(x-a)]^q} = \lim_{N \to \infty} \left\{ \left[\frac{N}{x-a}\right]^q \sum_{j=0}^{N-1} \frac{\Gamma(j-q)}{\Gamma(-q)\Gamma(j+1)} \left[\frac{Nx - jx + ja}{N} - a\right]\right\}$$

$$= [x-a]^{1-q} \left[\lim_{N \to \infty} \left\{ N^q \sum_{j=0}^{N-1} \frac{\Gamma(j-q)}{\Gamma(-q)\Gamma(j+1)}\right\}\right.$$

$$\left. - \lim_{N \to \infty} \left\{ N^{q-1} \sum_{j=0}^{N-1} j \frac{\Gamma(j-q)}{\Gamma(-q)\Gamma(j+1)}\right\}\right].$$

If the summation formulas (1.3.18) and (1.3.19) are now employed, followed by (1.3.14), the result
$$\frac{d^q[x-a]}{[d(x-a)]^q} = [x-a]^{1-q}\left[\frac{1}{\Gamma(1-q)} + \frac{q}{\Gamma(2-q)}\right]$$
is obtained which, on application of the recurrence formula (1.3.2), becomes

(4.3.1) $$\frac{d^q[x-a]}{[d(x-a)]^q} = \frac{[x-a]^{1-q}}{\Gamma(2-q)}.$$

Alternatively, we argue from the Riemann–Liouville formula (3.6.2) that, on substituting $w \equiv x - y$,

(4.3.2) $$\frac{d^q[x-a]}{[d(x-a)]^q} = \frac{1}{\Gamma(-q)} \int_a^x \frac{[y-a]\,dy}{[x-y]^{q+1}}$$
$$= \frac{1}{\Gamma(-q)} \int_0^{x-a} \frac{[x-a-w]\,dw}{w^{q+1}}$$
$$= \frac{1}{\Gamma(-q)} \left[\int_0^{x-a} \frac{[x-a]\,dw}{w^{q+1}} - \int_0^{x-a} \frac{dw}{w^q}\right]$$
$$= \frac{1}{\Gamma(-q)} \left[\frac{[x-a]^{1-q}}{-q} - \frac{[x-a]^{1-q}}{1-q}\right]$$
$$= \frac{[x-a]^{1-q}}{[-q][1-q]\Gamma(-q)}, \quad q < 0,$$

the denominator of which equals $\Gamma(2-q)$ by the recurrence formula (1.3.2). Use of equation (3.2.5),
$$\frac{d^q[x-a]}{[d(x-a)]^q} = \frac{d^n}{dx^n}\left\{\frac{d^{q-n}[x-a]}{[d(x-a)]^{q-n}}\right\},$$
is now all that is required to remove the restrictive q condition. Thus for arbitrary q one may select an integer n so large that $q - n < 0$. Result (4.3.2) gives
$$\frac{d^{q-n}[x-a]}{[d(x-a)]^{q-n}} = \frac{[x-a]^{1-q+n}}{\Gamma(2-q+n)},$$
whence
$$\frac{d^q[x-a]}{[d(x-a)]^q} = \frac{d^n}{[d(x-a)]^n}\left\{\frac{d^{q-n}[x-a]}{[d(x-a)]^{q-n}}\right\}$$
$$= \frac{\Gamma(2-q+n)}{\Gamma(2-q)} \frac{[x-a]^{1-q}}{\Gamma(2-q+n)} = \frac{[x-a]^{1-q}}{\Gamma(2-q)}$$
follows by equation (2.9.1).

We note, as expected, that formula (4.3.1) reduces to zero when $q = 2, 3, 4, \ldots$; to unity when $q = 1$; to $x - a$ when $q = 0$; and to $[x - a]^{n+1}/(n + 1)!$ when $q = -n = -1, -2, -3, \ldots$. Notice also, on comparison of formulas (4.3.1) and (4.1.1), that the qth differintegral of $x - a$ equals the $(q - 1)$th differintegral of unity. For this reason, Fig. 4.1.1 can be readily adapted to illustrate the present section.

4.4 THE FUNCTION $[x - a]^p$

The final function we consider in this chapter is the important function $f = [x - a]^p$, where p is initially arbitrary. We shall see, however, that p must exceed -1 for differintegration to have the properties we demand of the operator. (Recall our discussion of differintegrable functions in Section 3.1.)

For integer q of either sign, we have the formula (2.9.1) from classical calculus. Our first encounter with noninteger q will be restricted to negative q so that we may exploit the Riemann–Liouville definition; thus

$$\frac{d^q[x-a]^p}{[d(x-a)]^q} = \frac{1}{\Gamma(-q)} \int_a^x \frac{[y-a]^p \, dy}{[x-y]^{q+1}} = \frac{1}{\Gamma(-q)} \int_0^{x-a} \frac{v^p \, dv}{[x-a-v]^{q+1}}, \quad q < 0,$$

where v has replaced $y - a$. By further replacement of v by $[x - a]u$, the integral may be cast into the standard beta function form,

(4.4.1) $$\frac{d^q[x-a]^p}{[d(x-a)]^q} = \frac{[x-a]^{p-q}}{\Gamma(-q)} \int_0^1 u^p[1-u]^{-q-1} \, du, \quad q < 0.$$

The definite integral in (4.4.1) will be recognized as the beta function (see Section 1.3), $B(p + 1, -q)$, provided both arguments are positive. Therefore

(4.4.2) $$\frac{d^q[x-a]^p}{[d(x-a)]^q} = \frac{[x-a]^{p-q}}{\Gamma(-q)} B(p+1, -q) = \frac{\Gamma(p+1)[x-a]^{p-q}}{\Gamma(p-q+1)},$$

$$q < 1, \quad p > -1,$$

where the beta function has been replaced by its gamma function equivalent [see equation (1.3.24)].

For comparison of technique we digress at this point to include a verification of formula (4.4.2) starting with the definition (3.4.3). We replace $x - a$ by z so that

$$\frac{d^q[x-a]^p}{[d(x-a)]^q} \equiv \frac{d^q z^p}{dz^q} = \frac{\Gamma(q+1)}{2\pi i} \oint_c \frac{\zeta^p \, d\zeta}{[\zeta - z]^{q+1}},$$

where the contour C in the complex ζ-plane begins and ends at $\zeta = 0$ enclosing z once in the positive sense. If one sets $\zeta \equiv zs$, then

$$\frac{d^q z^p}{dz^q} = \frac{\Gamma(q+1)z^{p-q}}{2\pi i} \oint s^p[s-1]^{-q-1} \, ds,$$

where the integral is over a contour encircling the point $s = 1$ once in the positive sense and beginning and ending at $s = 0$. When such a contour is deformed into the one shown in Fig. 4.4.1, then

$$\begin{aligned}
\frac{d^q z^p}{dz^q} &= \frac{\Gamma(q+1)z^{p-q}}{2\pi i} \left[1 - \exp(-2\pi i[q+1])\right] \int_0^1 s^p[s-1]^{-q-1} \, ds \\
&= \frac{\Gamma(q+1)z^{p-q}}{2\pi i} \left[1 - \exp(-2\pi i[q+1])\right] [-]^{-q-1} \int_0^1 s^p[1-s]^{-q-1} \, ds \\
&= \frac{\Gamma(q+1)z^{p-q}}{2\pi i} \left[\exp(i\pi[q+1]) - \exp(-i\pi[q+1])\right] \int_0^1 s^p[1-s]^{-q-1} \, ds \\
&= \frac{\Gamma(q+1)z^{p-q}}{2\pi i} 2i \sin(\pi[q+1]) \int_0^1 s^p[1-s]^{-q-1} \, ds \\
&= \frac{\Gamma(p+1)z^{p-q}}{\Gamma(p-q+1)}, \qquad p > -1, \quad q < 0,
\end{aligned}$$

where use has been made of the reflection formula (1.3.8) and of the properties of the beta integral.

As in the previous section we may again use equation (3.2.5), together with the classical formula (2.9.1), to extend our treatment to positive q.

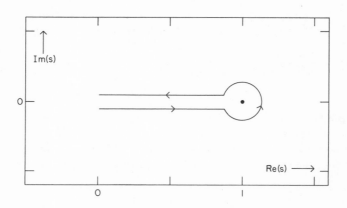

FIG. 4.4.1. Path of the contour for integration around the point $s = 1$.

4.4 THE FUNCTION $[x-a]^p$

Following this technique

(4.4.3)
$$\frac{d^q[x-a]^p}{[d(x-a)]^q} = \frac{d^n}{dx^n}\left[\frac{d^{q-n}[x-a]^p}{[d(x-a)]^{q-n}}\right]$$

$$= \frac{d^n}{dx^n}\left[\frac{[x-a]^{p-q+n}}{\Gamma(n-q)}\int_0^1 u^p[1-u]^{n-q-1}\,du\right]$$

$$= \frac{d^n}{dx^n}\left[\frac{\Gamma(p+1)[x-a]^{p-q+n}}{\Gamma(p-q+n+1)}\right], \qquad p > -1,$$

where, since we chose $n > q \geqq 0$, we were able to use (4.4.1) to evaluate the $[q-n]$th differintegral of $[x-a]^p$. The classical formula (2.9.1) then leads to

$$\frac{d^q[x-a]^p}{[d(x-a)]^q} = \frac{\Gamma(p+1)[x-a]^{p-q}}{\Gamma(p-q+1)}, \qquad q \geqq 0, \quad p > -1,$$

straightforwardly. Unification of this result with (4.4.2) yields the formula

(4.4.4)
$$\frac{d^q[x-a]^p}{[d(x-a)]^q} = \frac{\Gamma(p+1)[x-a]^{p-q}}{\Gamma(p-q+1)}, \qquad p > -1,$$

valid for all q. As required for an acceptable formula in our generalized calculus, equation (4.4.4) incorporates the classical formula (2.9.3).

Historically, the formula

$$\frac{d^q x^p}{dx^q} = \frac{\Gamma(p+1)x^{p-q}}{\Gamma(p-q+1)}$$

was important in being the basis of the concept of fractional differentiation as developed by Gemant (1936). This formulation was used by him, and later more extensively by Scott Blair et al. (1947) in rheology.

Thus far this section has been concerned only with the $p > -1$ instances of $[x-a]^p$. We now briefly deal with $p \leqq -1$. The generalized derivatives (4.4.1) and (4.4.3) break down for $p \leqq -1$ because the beta integrals then diverge. An infinite result

$$\frac{d^q[x-a]^p}{[d(x-a)]^q} = \infty, \qquad p \leqq -1, \text{ all } q,$$

would, however, be unacceptable because it would fail to incorporate the classical result (2.9.1) for positive integer q. Likewise, the formula (4.4.4) cannot be extended to $p \leqq -1$ because, though this does incorporate (2.9.1) it does not reproduce (2.9.2) for negative integer q. Moreover, we know of no generalization of formula (4.4.4) that incorporates both of the requirements (2.9.1) and (2.9.2) for $p \leqq -1$. The breakdown of (4.4.4) for $p \leqq -1$

4 DIFFERINTEGRATION OF SIMPLE FUNCTIONS

is associated with the pole of order unity or greater which occurs at $x = a$ for the functions $[x - a]^p$, $p \leq -1$. Functions for which such a pole occurs anywhere on the open interval from a to x lead to similar difficulties and for reasons such as this we have purposely excluded these functions from the class of differintegrable series, as explained in Section 3.1.

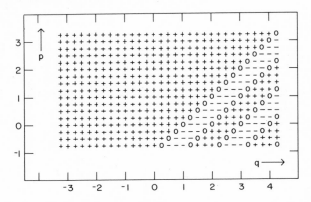

FIG. 4.4.2. The sign of the coefficient $\Gamma(p+1)/\Gamma(p-q+1)$ in the differintegrals $d^q[x-a]^p/[d(x-a)]^q$ for ranges of p and q values. Notice that entries stop short of the $p = -1$ line because functions $[x-a]^p$ are not differintegrable if $p \leq -1$.

Figure 4.4.2 shows the sign of the gamma function ratio that is the coefficient in the differintegral $d^q[x-a]^p/[d(x-a)]^q$ for ranges of values of p and q. Note that entries stop short of the $p = -1$ line and that there exist no conflicts between this diagram and Fig. 2.9.1.

CHAPTER 5
GENERAL PROPERTIES

In this chapter we examine those properties of differintegral operators which we might expect to generalize classical formulas for derivatives and integrals. It is these properties which will provide our primary means of understanding and utilizing the fractional calculus. As we shall see, while some of these classical properties do generalize without essential change, others require modification of one sort or another. Unless a stipulation is made to the contrary, we assume throughout the rest of the book that all functions encountered are differintegrable in the sense explained in Section 3.1. In the present chapter we shall often find it useful further to restrict our attention to differintegrable series.

5.1 LINEARITY

The linearity of the differintegral operator, by which we mean

$$(5.1.1) \qquad \frac{d^q[f_1 + f_2]}{[d(x-a)]^q} = \frac{d^q f_1}{[d(x-a)]^q} + \frac{d^q f_2}{[d(x-a)]^q},$$

is an immediate consequence of any of the definitions summarized in Section 3.6.

5.2 DIFFERINTEGRATION TERM BY TERM

The linearity of differintegral operators means that they may be distributed through the terms of a finite sum; i.e.,

$$(5.2.1) \qquad \frac{d^q}{[d(x-a)]^q} \sum_{j=0}^{n} f_j = \sum_{j=0}^{n} \frac{d^q f_j}{[d(x-a)]^q}.$$

5 GENERAL PROPERTIES

We want to investigate the circumstances that permit term-by-term differintegration of an infinite series of functions. Our major goal in this section is to establish the term-by-term differintegrability of general differintegrable series (see Section 3.1). We shall make repeated use of the classical results on differentiation and integration of infinite series term by term (Section 2.8). In order to use these results we must ensure that the terms f_j of the series are either continuous or continuously differentiable. If we restrict our attention to summands f_j that are differintegrable series, the form of such a series [see equation (3.1.1)] and of its term-by-term derivative shows that the requisite continuity assumptions are valid away from the lower limit $x = a$. Henceforth we shall consider infinite sums of differintegrable series and establish results on the term-by-term differintegrability of such sums which will, in general, be valid in open intervals[1] such as $a < x < a + X$, where X is the radius of convergence of the differintegrable series. First we need to establish some facts about this radius of convergence.

Consider first an ordinary power series,

$$\phi = \sum_{j=0}^{\infty} a_j [x - a]^j, \qquad a_j = \frac{\phi^{(j)}(a)}{j!},$$

convergent for $|x - a| \leq X$. One knows from classical results that ϕ, together with all of its term-by-term derivatives and integrals, converges uniformly in the interval $0 \leq |x - a| < X$. What can be said about the series obtained, more generally, from term-by-term differintegration of ϕ? Making use of equation (4.4.4), the series obtained by applying $d^q/[d(x-a)]^q$ to every summand of ϕ is the series

$$\sum_{j=0}^{\infty} \frac{a_j \Gamma(j+1)}{\Gamma(j-q+1)} [x-a]^{j-q} = [x-a]^{-q} \sum_{j=0}^{\infty} \frac{\phi^{(j)}(a)}{\Gamma(j-q+1)} [x-a]^j.$$

We know that the series for ϕ converges for $|x - a| < X$, where, by the ratio test,

$$X \equiv \lim_{j \to \infty} \left| \frac{a_j}{a_{j+1}} \right| = \lim_{j \to \infty} \left| \frac{[j+1]\phi^{(j)}(a)}{\phi^{(j+1)}(a)} \right|,$$

while the differintegrated series will converge for

$$|x - a| < \lim_{j \to \infty} \left| \frac{\phi^{(j)}(a)\Gamma(j-q+2)}{\phi^{(j+1)}(a)\Gamma(j-q+1)} \right| = \lim_{j \to \infty} \left| \frac{[j-q+1]\phi^{(j)}(a)}{\phi^{(j+1)}(a)} \right|$$

$$\geq \lim_{j \to \infty} \left| \frac{[j+1]\phi^{(j)}(a)}{\phi^{(j+1)}(a)} \right| - q \lim_{j \to \infty} \left| \frac{\phi^{(j)}(a)}{\phi^{(j+1)}(a)} \right|$$

$$= X - qA,$$

---

[1] The classical results of Section 2.8 are stated for closed intervals rather than open ones. Henceforth the phrase "convergence in an open interval" is adopted as a shorthand for "convergence in every closed subinterval of the open interval."

5.2 DIFFERINTEGRATION TERM BY TERM   71

where

$$A \equiv \lim_{j \to \infty} \left| \frac{\phi^{(j)}(a)}{\phi^{(j+1)}(a)} \right|.$$

Examination of the possibilities for A and X reveals that A is invariably negligible in comparison with X so that the differintegrated series converges in the open interval $0 < |x - a| < X$. Furthermore, the same result is valid for the differintegrable series whose jth term is $a_j[x - a]^{j+p}$ (since this series has the same radius of convergence as its analytic part, $\sum a_j[x - a]^j$) and, therefore, for general differintegrable series. That is, if the differintegrable series f, which is a finite sum of functions each representable as

$$[x - a]^p \sum_{j_1=0}^{\infty} a_{j_1}[x - a]^{j_1} + [x - a]^{[np+1]/n} \sum_{j_2=0}^{\infty} a_{j_2}[x - a]^{j_2}$$

$$+ \cdots + [x - a]^{[np+n-1]/n} \sum_{j_n=0}^{\infty} a_{j_n}[x - a]^{j_n},$$

converges for $|x - a| < X$, then so does the series obtained by differintegrating each "unit" term by term, except possibly at the endpoint $x = a$. This fact will be important in what follows.

Let f be any differintegrable series. Since f may be decomposed as a finite sum of differintegrable series units

$$f_U = [x - a]^p \sum_{j=0}^{\infty} a_j[x - a]^j,$$

where $p > -1$ and $a_0 \neq 0$, the term-by-term differintegrability of f will follow from that of f_U. Accordingly, our objective is to establish that

$$(5.2.2) \qquad \frac{d^q}{[d(x - a)]^q} \left\{ [x - a]^p \sum_{j=0}^{\infty} a_j[x - a]^j \right\} = \sum_{j=0}^{\infty} a_j \frac{d^q[x - a]^{p+j}}{[d(x - a)]^q}$$

for all q. More specifically, the equality (5.2.2) will be proven valid inside the interval of convergence of the differintegrable series $\sum a_j[x - a]^{p+j}$.

For $q \leq 0$ a stronger result that directly extends the classical theorem on term-by-term integration is easy to establish: Suppose the infinite series of differintegrable functions $\sum f_j$ converges uniformly in $0 < |x - a| < X$; then

$$(5.2.3) \qquad \frac{d^q}{[d(x - a)]^q} \sum_{j=0}^{\infty} f_j = \sum_{j=0}^{\infty} \frac{d^q f_j}{[d(x - a)]^q}, \qquad q \leq 0,$$

and the right-hand series also converges uniformly in $0 < |x - a| < X$. To demonstrate this result, let

$$f \equiv \sum_{j=0}^{\infty} f_j, \qquad S_N \equiv \sum_{j=0}^{N} f_j.$$

5 GENERAL PROPERTIES

Since $q < 0$, the Riemann–Liouville representations

$$\frac{d^q f}{[d(x-a)]^q} = \frac{1}{\Gamma(-q)} \int_a^x \frac{f(y)\,dy}{[x-y]^{q+1}}, \qquad \frac{d^q f_j}{[d(x-a)]^q} = \frac{1}{\Gamma(-q)} \int_a^x \frac{f_j(y)\,dy}{[x-y]^{q+1}}$$

are valid and

$$\frac{d^q f}{[d(x-a)]^q} - \frac{d^q S_N}{[d(x-a)]^q} = \frac{1}{\Gamma(-q)} \int_a^x \frac{[f(y) - S_N(y)]\,dy}{[x-y]^{q+1}}.$$

The assumption of uniform convergence means that, given $\varepsilon > 0$, there is an integer $N = N(\varepsilon)$ such that

$$|f(y) - S_n(y)| < \varepsilon$$

for $n > N$ and for all y in the interval $a \leq y \leq x$ with $|x - a| < X$. Then

$$\left| \frac{d^q f}{[d(x-a)]^q} - \sum_{j=0}^N \frac{d^q f_j}{[d(x-a)]^q} \right| = \frac{1}{\Gamma(-q)} \left| \int_a^x \frac{[f(y) - S_N(y)]\,dy}{[x-y]^{q+1}} \right|$$

$$\leq \frac{1}{\Gamma(-q)} \int_a^x \frac{|f(y) - S_N(y)|\,dy}{[x-y]^{q+1}}$$

$$< \frac{\varepsilon}{\Gamma(-q)} \int_a^x [x-y]^{-q-1}\,dy$$

$$= \frac{\varepsilon[x-a]^{-q}}{q\Gamma(-q)},$$

which can be made small independently of x in the interval $0 < |x - a| < X$. This proves that $\sum d^q f_j / [d(x-a)]^q$ converges uniformly to $d^q f / [d(x-a)]^q$ in $0 < |x - a| < X$.

The result just established shows that equation (5.2.2) is valid for $q \leq 0$, and thus, if f is any differintegrable series, the operator $d^q/[d(x-a)]^q$ may be distributed through the several infinite series that define f as long as $q \leq 0$. Applying result (4.4.4) to equation (5.2.2) gives

$$(5.2.4) \qquad \frac{d^q f_U}{[d(x-a)]^q} = \sum_{j=0}^\infty \frac{a_j \Gamma(p+j+1)}{\Gamma(p-q+j+1)} [x-a]^{p+j-q}, \qquad q \leq 0.$$

Equations (5.2.2) and (5.2.4) are also valid for $q > 0$, as we now establish. First we decompose the series for f_U into two pieces,

$$f_U = \sum_{j=0}^\infty a_j [x-a]^{p+j} = \sum_{j \in J_1} a_j [x-a]^{p+j} + \sum_{j \in J_2} a_j [x-a]^{p+j},$$

where J_1 is the set of nonnegative integers j for which $\Gamma(p - q + j + 1)$ is infinite and J_2 consists of all nonnegative integers not in J_1. For fixed q

5.2 DIFFERINTEGRATION TERM BY TERM

the properties of the gamma function (Section 1.3) ensure that the set J_1 has only a finite number of elements. Thus,

$$\frac{d^q f_U}{[d(x-a)]^q} = \frac{d^q}{[d(x-a)]^q}\left\{\sum_{j\in J_1} a_j [x-a]^{p+j}\right\} + \frac{d^q}{[d(x-a)]^q}\left\{\sum_{j\in J_2} a_j [x-a]^{p+j}\right\}$$

$$= \sum_{j\in J_1} \frac{d^q [x-a]^{p+j}}{[d(x-a)]^q} + \frac{d^q}{[d(x-a)]^q}\left\{\sum_{j\in J_2} a_j [x-a]^{p+j}\right\},$$

making use only of the linearity of $d^q/[d(x-a)]^q$. Now we see that the proof of equation (5.2.4) for $q > 0$ depends only upon establishing that

(5.2.5)
$$\frac{d^q}{[d(x-a)]^q}\left\{\sum_{j\in J_2} a_j [x-a]^{p+j}\right\} = \sum_{j\in J_2} a_j \frac{d^q [x-a]^{p+j}}{[d(x-a)]^q}$$

$$= \sum_{j\in J_2} \frac{a_j \Gamma(p+j+1)}{\Gamma(p-q+j+1)} [x-a]^{p+j-q}, q > 0.$$

Assuming that the series for f_U converges uniformly in $0 < |x-a| < X$, so will the series on the right-hand side of equation (5.2.5), as we proved at the beginning of this section. Thus the operator $d^{-1}/[d(x-a)]^{-1}$ may be distributed through the terms of this series to yield

(5.2.6) $$\frac{d^{-1}}{[d(x-a)]^{-1}}\left\{\sum_{j\in J_2} \frac{a_j \Gamma(p+j+1)}{\Gamma(p-q+j+1)} [x-a]^{p+j-q}\right\}$$

$$= \sum_{j\in J_2} \frac{d^{-1}}{[d(x-a)]^{-1}} \left\{\frac{a_j \Gamma(p+j+1)}{\Gamma(p-q+j+1)} [x-a]^{p+j-q}\right\}$$

$$= \sum_{j\in J_2} \frac{a_j \Gamma(p+j+1)\Gamma(p-q+j+1)}{\Gamma(p-q+j+1)\Gamma(p-q+j+2)} [x-a]^{p+j-q+1}$$

$$= \sum_{j\in J_2} \frac{a_j \Gamma(p+j+1)}{\Gamma(p-q+j+2)} [x-a]^{p+j-q+1}$$

$$= \sum_{j\in J_2} a_j \frac{d^{q-1}}{[d(x-a)]^{q-1}} [x-a]^{p+j}$$

and the last series converges uniformly in $0 < |x-a| < X$, as does the series obtained from it by differentiating each term. The cancellation needed to obtain the penultimate expression in equation (5.2.6) may be justified since the definition of the set J_2 guarantees that $\Gamma(p-q+j+1)$ is finite. Applying

5 GENERAL PROPERTIES

the classical theorem on term-by-term differentiation (Section 2.8) to the series $\sum a_j \{d^{q-1}/[d(x-a)]^{q-1}\}[x-a]^{p+j}$ gives

$$\frac{d}{dx}\left\{\sum_{j\in J_2} a_j \frac{d^{q-1}}{[d(x-a)]^{q-1}}[x-a]^{p+j}\right\} = \sum_{j\in J_2} a_j \frac{d^q}{[d(x-a)]^q}[x-a]^{p+j}.$$

Arguing similarly we find that

(5.2.7) $\quad \dfrac{d^n}{dx^n}\left\{\sum_{j\in J_2} a_j \dfrac{d^{q-n}}{[d(x-a)]^{q-n}}[x-a]^{p+j}\right\} = \sum_{j\in J_2} a_j \dfrac{d^q}{[d(x-a)]^q}[x-a]^{p+j}$

for every positive integer n. Choosing n to make $q - n < 0$ permits us to apply equation (5.2.4), with the result that

$$\sum_{j\in J_2} a_j \frac{d^{q-n}}{[d(x-a)]^{q-n}}[x-a]^{p+j} = \frac{d^{q-n}}{[d(x-a)]^{q-n}}\left\{\sum_{j\in J_2} a_j[x-a]^{p+j}\right\}.$$

Differentiating both sides of this equation n times, we see that

$$\frac{d^n}{dx^n}\sum_{j\in J_2} a_j \frac{d^{q-n}}{[d(x-a)]^{q-n}}[x-a]^{p+j} = \frac{d^q}{[d(x-a)]^q}\left\{\sum_{j\in J_2} a_j[x-a]^{p+j}\right\}.$$

Utilizing equation (5.2.7) gives, finally,

$$\frac{d^q}{[d(x-a)]^q}\left\{\sum_{j\in J_2} a_j[x-a]^{p+j}\right\} = \sum_{j\in J_2} a_j \frac{d^q}{[d(x-a)]^q}[x-a]^{p+j}, \qquad q > 0,$$

as we wanted to show. Thus the representation (5.2.5) is valid for $q > 0$ and, hence, for arbitrary q.

We have accomplished our principal goal in this section: to prove the term-by-term differintegrability of arbitrary differintegrable series. In the process we established a generalization (to the operator $d^q/[d(x-a)]^q$ for any $q \leq 0$) of the classical theorem on term-by-term integration, valid for uniformly convergent series. One may wonder whether a similar generalization to $q > 0$ of the classical theorem on term-by-term differentiation is valid. The answer is in the affirmative: If the infinite series $\sum f_j$ as well as the series $\sum d^q f_j / [d(x-a)]^q$ converge uniformly in $0 < |x - a| < X$, then

(5.2.8) $\qquad \dfrac{d^q}{[d(x-a)]^q} \displaystyle\sum_{j=0}^{\infty} f_j = \sum_{j=0}^{\infty} \dfrac{d^q f_j}{[d(x-a)]^q}, \qquad q > 0,$

for $0 < |x - a| < X$. The proof of this result makes use of composition rule facts not developed fully until much later (see Sections 5.7 and 8.4) and will be omitted. We note here only that the result does provide a very natural extension of the classical theorem on term-by-term differentiation (Section 2.8) which also required, in addition to convergence of $\sum f_j$, the uniform convergence of $\sum df_j/dx$.

Equation (5.2.4) provides a useful alternative representation for the differintegral of an analytic function

$$\text{(5.2.9)} \qquad \frac{d^q\phi}{[d(x-a)]^q} = \sum_{j=0}^{\infty} \frac{\phi^{(j)}(a)}{\Gamma(j-q+1)} [x-a]^{j-q},$$

where

$$\phi = \sum_{j=0}^{\infty} \frac{\phi^{(j)}(a)}{\Gamma(j+1)} [x-a]^j.$$

The series in (5.2.9) converges uniformly in $0 < |x-a| < X$, where X is the radius of convergence of ϕ. Equation (5.2.9) was previously presented without proof as formula (3.6.6).

5.3 HOMOGENEITY

The proof of homogeneity,

$$\text{(5.3.1)} \qquad \frac{d^q[Cf]}{[d(x-a)]^q} = C \frac{d^q f}{[d(x-a)]^q}, \qquad C \text{ any constant,}$$

follows directly from the definition (3.6.1) since the constant C may be brought outside the sum and limit.

5.4 SCALE CHANGE

By a scale change of the function f with respect to a lower limit a, we mean its replacement by $f(\beta x - \beta a + a)$, where β is a constant termed the scaling factor. To clarify this definition, consider $a = 0$; then the scale change converts $f(x)$ to $f(\beta x)$, in contrast to the homogeneity operation of the previous section which converted $f(x)$ to $Cf(x)$.

In this section we seek a procedure by which the effect of the generalized $d^q/[d(x-a)]^q$ operation upon $f(\beta x - \beta a + a)$ can be found, if $d^q f/[d(x-a)]^q$ is known. We shall find it convenient to use the abbreviation

$$X \equiv x + [a - a\beta]/\beta$$

and to adopt the Riemann–Liouville definition (3.6.2). Using Y as a replacement for $\beta y - \beta a + a$, we proceed as follows:

$$(5.4.1) \quad \frac{d^q f(\beta X)}{[d(x-a)]^q} = \frac{d^q f(\beta x - \beta a + a)}{[d(x-a)]^q} = \frac{1}{\Gamma(-q)} \int_a^x \frac{f(\beta y - \beta a + a)\,dy}{[x-y]^{q+1}}$$

$$= \frac{1}{\Gamma(-q)} \int_a^{\beta X} \frac{f(Y)[dY/\beta]}{\{[\beta X - Y]/\beta\}^{q+1}} = \frac{\beta^q}{\Gamma(-q)} \int_a^{\beta X} \frac{f(Y)\,dY}{[\beta X - Y]^{q+1}}$$

$$= \beta^q \frac{d^q f(\beta X)}{[d(\beta X - a)]^q}.$$

The utility of formula (5.4.1) is greatest when $a = 0$, for then $X = x$ and the scale change is simply a multiplication of the independent variable by a constant, the formula being

$$(5.4.2) \quad \frac{d^q f(\beta x)}{[dx]^q} = \beta^q \frac{d^q f(\beta x)}{[d(\beta x)]^q}.$$

This result may also be found in the work of Erdélyi *et al.* (1954). When a is nonzero, the effect of replacing $f(x)$ by $f(\beta x)$ requires, in addition to a scale change, the algebraically more difficult translation process, consideration of which will be deferred until Section 5.9.

5.5 LEIBNIZ'S RULE

The rule for differentiation of a product of two functions is a familar result in elementary calculus. It states that

$$(5.5.1) \quad \frac{d^n[fg]}{dx^n} = \sum_{j=0}^n \binom{n}{j} \frac{d^{n-j}f}{dx^{n-j}} \frac{d^j g}{dx^j}$$

and is, of course, restricted to nonnegative integers n. In Section 2.5 we have derived, based on integration by parts, the following product rule for multiple integrals:

$$\frac{d^{-n}[fg]}{[d(x-a)]^{-n}} = \sum_{j=0}^\infty \binom{-n}{j} \frac{d^{-n-j}f}{[d(x-a)]^{-n-j}} \frac{d^j g}{[d(x-a)]^j}.$$

When we observe that the finite sum in (5.5.1) could equally well extend to infinity [since $\binom{n}{j} = 0$ for $j > n$] we might expect the product rule to generalize to arbitrary order q as

$$(5.5.2) \quad \frac{d^q[fg]}{[d(x-a)]^q} = \sum_{j=0}^\infty \binom{q}{j} \frac{d^{q-j}f}{[d(x-a)]^{q-j}} \frac{d^j g}{[d(x-a)]^j}.$$

5.5 LEIBNIZ'S RULE

That such a generalization is indeed valid for real analytic functions $\phi(x)$ and $\psi(x)$ will now be established.

Starting with equation (4.1.2) and substituting for ϕ the product $\phi\psi$, we obtain

$$\frac{d^q[\phi\psi]}{[d(x-a)]^q} = \sum_{k=0}^{\infty} \binom{q}{k} \frac{d^{q-k}[1]}{[d(x-a)]^{q-k}} [\phi\psi]^{(k)}$$

$$= \sum_{k=0}^{\infty} \binom{q}{k} \frac{d^{q-k}[1]}{[d(x-a)]^{q-k}} \sum_{j=0}^{k} \binom{k}{j} \phi^{(k-j)} \psi^{(j)},$$

making use of (5.5.1). Note that, since j is an integer, the repeated derivative $\psi^{(j)}$ with respect to x equals that with respect to $x - a$. The permutation (see footnote 5 in Section 2.5)

(5.5.3) $$\sum_{k=0}^{\infty} \sum_{j=0}^{k} = \sum_{j=0}^{\infty} \sum_{k=j}^{\infty}$$

may be applied to give

$$\frac{d^q[\phi\psi]}{[d(x-a)]^q} = \sum_{j=0}^{\infty} \psi^{(j)} \sum_{k=j}^{\infty} \binom{q}{k}\binom{k}{j} \frac{d^{q-k}[1]}{[d(x-a)]^{q-k}} \phi^{(k-j)}$$

$$= \sum_{j=0}^{\infty} \psi^{(j)} \sum_{l=0}^{\infty} \binom{q}{l+j}\binom{l+j}{j} \frac{d^{q-j-l}[1]}{[d(x-a)]^{q-j-l}} \phi^{(l)}$$

$$= \sum_{j=0}^{\infty} \binom{q}{j} \psi^{(j)} \sum_{l=0}^{\infty} \binom{q-j}{l} \frac{d^{q-j-l}[1]}{[d(x-a)]^{q-j-l}} \phi^{(l)}$$

$$= \sum_{j=0}^{\infty} \binom{q}{j} \psi^{(j)} \frac{d^{q-j}\phi}{[d(x-a)]^{q-j}},$$

where we have made use of the identity[2]

$$\binom{q}{l+j}\binom{l+j}{j} = \binom{q}{j}\binom{q-j}{l}$$

and a second application of (4.1.2). Since we established (3.5.3) under the assumption that ϕ is a real analytic function and used (3.5.3) to prove (4.1.2), which in turn was used to establish (5.5.2), the latter is proven only if ϕ and ψ are real analytic functions.

Nevertheless, a somewhat different argument may be used to establish equation (5.5.2) when one of the functions is a polynomial. The argument

---

[2] This identity is immediately apparent when the binomial coefficients are replaced by gamma function equivalents.

5 GENERAL PROPERTIES

begins with consideration of the product $xf(x)$ and $q < 0$. Making use of the Riemann-Liouville definition,

$$(5.5.4) \quad \frac{d^q[xf]}{[d(x-a)]^q} = \frac{1}{\Gamma(-q)} \int_a^x \frac{yf(y)\,dy}{[x-y]^{q+1}} + \frac{x}{\Gamma(-q)} \int_a^x \frac{f(y)\,dy}{[x-y]^{q+1}}$$

$$- \frac{x}{\Gamma(-q)} \int_a^x \frac{f(y)\,dy}{[x-y]^{q+1}}$$

$$= \frac{x}{\Gamma(-q)} \int_a^x \frac{f(y)\,dy}{[x-y]^{q+1}} - \frac{1}{\Gamma(-q)} \int_a^x \frac{f(y)\,dy}{[x-y]^q}$$

$$= x \frac{d^q f}{[d(x-a)]^q} + q \frac{d^{q-1} f}{[d(x-a)]^{q-1}},$$

where, in the last step, use was made of the recurrence formula (1.3.2) for the gamma function. Extension of this result to $q \geq 0$ is now quite easy, for if $n - 1 \leq q < n$, $n = 1, 2, 3, \ldots$, then

$$\frac{d^q[xf]}{[d(x-a)]^q} = \frac{d^n}{dx^n} \left\{ \frac{d^{q-n}[xf]}{[d(x-a)]^{q-n}} \right\}$$

$$= \frac{d^n}{dx^n} \left\{ x \frac{d^{q-n}f}{[d(x-a)]^{q-n}} + [q-n] \frac{d^{q-n-1}f}{[d(x-a)]^{q-n-1}} \right\}$$

$$= x \frac{d^q f}{[d(x-a)]^q} + n \frac{d^{q-1}f}{[d(x-a)]^{q-1}} + [q-n] \frac{d^{q-1}f}{[d(x-a)]^{q-1}}$$

$$= x \frac{d^q f}{[d(x-a)]^q} + q \frac{d^{q-1}f}{[d(x-a)]^{q-1}}$$

as before. Equation (3.2.5) has been used repeatedly in the preceding derivation. An inductive argument establishes equation (5.5.2) when $g = x^k$ for any nonnegative integer k and any f, thus, for g any polynomial and f arbitrary.

When g is a polynomial in $x - a$ in equation (5.5.2) the sum on the right-hand side is, of course, finite. Convergence difficulties are encountered, however, when one tries to extend the previous argument to the case f arbitrary and g analytic. An example that illustrates this is obtained by considering $f = 1$, $g = [1-x]^{-1} = 1 + x + x^2 + \cdots$ for $|x| < 1$. Then

$$\frac{d^q[fg]}{dx^q} = \frac{d^q\{[1-x]^{-1}\}}{dx^q} = \frac{d^q}{dx^q} \sum_{k=0}^\infty x^k, \qquad |x| < 1.$$

Term-by-term differintegration is justified by the results of Section 5.2 since

$\sum x^k$ is certainly a differintegrable (in fact, an analytic) series. Therefore,
$$\frac{d^q}{dx^q}\sum_{k=0}^{\infty} x^k = \sum_{k=0}^{\infty} \frac{d^q x^k}{dx^q} = \sum_{k=0}^{\infty} \frac{\Gamma(k+1)}{\Gamma(k-q+1)} x^{k-q},$$
which converges for $0 < |x| < 1$. On the other hand, use of Leibniz's rule would give
$$\frac{d^q}{dx^q}\sum_{k=0}^{\infty} x^k = \sum_{j=0}^{\infty} \binom{q}{j} \frac{d^{q-j}[1]}{dx^{q-j}} \frac{d^j\{[1-x]^{-1}\}}{dx^j}$$
$$= \sum_{j=0}^{\infty} \frac{\Gamma(q+1)}{\Gamma(q-j+1)\Gamma(j+1)} \frac{x^{j-q}}{\Gamma(j-q+1)} \Gamma(j+1)[1-x]^{-j-1}$$
$$= \frac{\Gamma(q+1)x^{-q}}{1-x} \sum_{j=0}^{\infty} \frac{[-]^j}{[j-q]\pi\csc(\pi q)} \frac{x^j}{[1-x]^j},$$
which converges only for $x \leqq \frac{1}{2}$.

Leibniz's rule has been thoroughly studied recently by Osler (1970a, 1971, 1972b, 1972c). He was led to wonder whether equation (5.5.2) is a special case of a still more general result in which the interchangeability of f and g is more apparent. The more general result proved by Osler is

(5.5.5) $$\frac{d^q[fg]}{dx^q} = \sum_{j=-\infty}^{\infty} \frac{\Gamma(q+1)}{\Gamma(q-\gamma-j+1)\Gamma(\gamma+j+1)} \frac{d^{q-\gamma-j}f}{dx^{q-\gamma-j}} \frac{d^{\gamma+j}g}{dx^{\gamma+j}},$$

where γ is arbitrary, which reduces to (5.5.2) with $a = 0$ when $\gamma = 0$. Watanabe derived equation (5.5.5) in 1931, but his method does not yield the precise region of convergence in the complex plane. Osler pointed this out in his paper (1970a) and used the Cauchy integral formula representation [epuation (3.4.3)] for fractional derivatives to delineate the appropriate region of convergence (a star-shaped region in the complex plane) for the formula (5.5.5). He also used (5.5.5) to generate certain infinite series expansions interrelating special functions of mathematical physics. The interested reader is referred to the work of Osler (1970a, 1972b) for further details.

A further generalization of Leibniz's rule due to Osler (1972c) is the integral form

(5.5.6) $$\frac{d^q[fg]}{dx^q} = \int_{-\infty}^{\infty} \frac{\Gamma(q+1)}{\Gamma(q-\gamma-\lambda+1)\Gamma(\gamma+\lambda+1)} \frac{d^{q-\gamma-\lambda}f}{dx^{q-\gamma-\lambda}} \frac{d^{\gamma+\lambda}g}{dx^{\gamma+\lambda}} d\lambda$$

in which a discrete sum is replaced by an integral. Formula (5.5.6) has been used by Osler to obtain a generalization of Parseval's integral formula [see Titchmarsh (1948)] of Fourier analysis. As one might expect, equation (5.5.6) is also useful in deriving formulas for definite integrals. A table of these is presented in Osler's work (1972c).

5.6 CHAIN RULE

The chain rule for differentiation,

$$\frac{d}{dx}g(f(x)) = \frac{d}{df(x)}g(f(x))\frac{d}{dx}f(x),$$

lacks a simple counterpart in the integral calculus. Indeed, if there were such a counterpart, the process of integration would pose no greater difficulty than does differentiation. Since any general formula for $d^q g(f(x))/[d(x-a)]^q$ must encompass integration as a special case, little hope can be held out for a useful chain rule for arbitrary q. Nevertheless, as we shall see, a formal chain rule may be derived quite simply.

We start with formula (4.1.2) valid for an analytic function ϕ:

$$\frac{d^q \phi}{[d(x-a)]^q} = \sum_{j=0}^{\infty} \binom{q}{j} \frac{d^{q-j}[1]}{[d(x-a)]^{q-j}} \frac{d^j \phi}{dx^j}.$$

The formula (4.1.1) permits the evaluation of the effect of the differintegral operator upon unity, allowing us to write

$$\frac{d^q \phi}{[d(x-a)]^q} = \frac{[x-a]^{-q}}{\Gamma(1-q)} \phi + \sum_{j=1}^{\infty} \binom{q}{j} \frac{[x-a]^{j-q}}{\Gamma(j-q+1)} \frac{d^j \phi}{dx^j},$$

where the $j=0$ term has been separated from the others. Up to this point we have regarded ϕ as a function of x. Now we consider $\phi = \phi(f(x))$ and evaluate $d^j \phi(f(x))/dx^j$ by Faà de Bruno's formula, as developed in Section 2.6:

$$\frac{d^q}{[d(x-a)]^q} \phi(f(x)) = \frac{[x-a]^{-q}}{\Gamma(1-q)} \phi(f(x))$$

$$+ \sum_{j=1}^{\infty} \binom{q}{j} \frac{[x-a]^{j-q}}{\Gamma(j-q+1)} j! \sum_{m=1}^{j} \phi^{(m)} \sum \prod_{k=1}^{j} \frac{1}{P_k!} \left[\frac{f^{(k)}}{k!}\right]^{P_k},$$

where the third summation and the P_k's have the significances explained in Section 2.6.

The complexity of this result will inhibit its general utility. We see on inserting $q = -1$ that even for the case of a single integration,

$$\int_a^x \phi(f(y)) \, dy = [x-a]\phi(f(x))$$

$$+ \sum_{j=1}^{\infty} [-]^j \frac{[x-a]^{j+1}}{j+1} \sum_{m=1}^{j} \phi^{(m)} \sum \prod_{k=1}^{j} \frac{1}{P_k!} \left[\frac{f^{(k)}}{k!}\right]^{P_k},$$

the chain rule gives an infinite series that offers little hope of being expressible in closed form, except for trivially simple instances of the functions f and ϕ.

A case in which the generalized chain rule could be of limited utility is provided by $f(x) = \exp(x)$. Then

$$\frac{d^q}{[d(x-a)]^q} \phi(\exp(x)) = \frac{[x-a]^{-q}}{\Gamma(1-q)} \phi(\exp(x))$$

$$+ \sum_{j=1}^{\infty} \binom{q}{j} \frac{[x-a]^{j-q}}{\Gamma(j-q+1)} \exp(jx) \sum_{m=1}^{j} S_j^{[m]} \phi^{(m)},$$

where $S_j^{[m]}$ is a Stirling number of the second kind (see the discussion in Section 1.3).

Extension of the chain rule in a somewhat different direction has recently been accomplished by Osler (1970b). For composite functions $f(x) = F(h(x))$ his aim was to generalize to fractional derivatives the classical formula

$$\frac{d^N f(x)}{dx^N} = \sum_{n=0}^{N} \frac{U_n(x)}{n!} \frac{d^n f(x)}{[dh(x)]^n},$$

where

$$U_n(x) = \sum_{r=0}^{n} \binom{n}{r} [-h(x)]^r \frac{d^N}{dx^N} h(x)^{n-r}.$$

He obtained such a generalization and, with the help of this and the generalized Leibniz rule (5.5.5), he derived some known and some apparently new relationships involving special functions of mathematical physics.

The absence of a simple chain rule impedes the process of differintegration with respect to a variable other than the argument of the differintegrand. In other words, considering here only those cases where the lower limit $a = 0$, it is difficult to relate

$$\frac{d^q}{dX^q} f(x) \quad \text{to} \quad \frac{d^q}{dx^q} f(x)$$

even where X is a very simple function of x. Of course if $X = \beta x$, where β is a constant, the scale change theorem of Section 5.4 does provide such a relationship.

Another useful case in which a relationship may be derived is the case $X = x^n$, n being a positive integer, and f being one of the many functions expressible as a generalized hypergeometric function (Section 2.10 and Chapter 9). In that event, we have

$$\frac{d^q}{dX^q} f(x) = \frac{d^q}{dX^q} \left[X^{1/n} \frac{b_1, b_2, \ldots, b_K}{c_1, c_2, \ldots, c_L} \right]$$

so that the differintegration may be carried out via the mediation of equation (2.10.4).

5.7 COMPOSITION RULE

In seeking a general composition rule for the operator $d^q/[d(x-a)]^q$ we search for the relationship between

$$\frac{d^q}{[d(x-a)]^q}\frac{d^Qf}{[d(x-a)]^Q} \quad \text{and} \quad \frac{d^{q+Q}f}{[d(x-a)]^{q+Q}},$$

which we temporarily abbreviate to $d^q d^Q f$ and $d^{q+Q}f$. Of course, if these symbols are to be generally meaningful we need to assume not only that f is differintegrable but that $d^Q f$ is differintegrable as well. In the present section we restrict attention to differintegrable series as defined in Section 3.1.

We saw in Section 3.1 that the most general nonzero differintegrable series is a finite sum of differintegrable "units," each having the form

$$(5.7.1) \qquad f_U = [x-a]^p \sum_{j=0}^{\infty} a_j [x-a]^j, \quad p > -1, \quad a_0 \neq 0.$$

We shall see that the composition rule may be valid for some units of f but possibly not for others. It follows from the linearity of differintegral operators that

$$(5.7.2) \qquad d^q d^Q f = d^{q+Q} f$$

if

$$(5.7.3) \qquad d^q d^Q f_U = d^{q+Q} f_U$$

for every unit f_U of f. Accordingly we shall first assess the validity of the composition rule (5.7.3) for a differintegrable series unit function f_U.

Obviously, if $f_U = 0$, then $d^Q f_U = 0$ for every Q by equation (4.2.2), and so

$$d^q d^Q [0] = d^{q+Q}[0] = 0.$$

While the composition rule is trivially satisfied for the differintegrable function $f_U = 0$, we shall see that the possibility

$$f_U \neq 0, \quad \text{but} \quad d^Q f_U = 0,$$

is exactly the condition that prevents the composition rule (5.7.3), and therefore (5.7.2), from being satisfied generally.

Having dealt with the case $f_U = 0$ we now assume $f_U \neq 0$ and use equation (5.2.4) to evaluate $d^Q f_U$:

$$(5.7.4) \qquad d^Q f_U = \sum_{j=0}^{\infty} a_j d^Q [x-a]^{p+j} = \sum_{j=0}^{\infty} \frac{a_j \Gamma(p+j+1)[x-a]^{p+j-Q}}{\Gamma(p+j-Q+1)}.$$

Furthermore, we note that since $p > -1$, it follows that $p + j > -1$ so that $\Gamma(p+j+1)$ is always finite but nonzero. Individual terms in $d^Q f_U$ will vanish, therefore, only when the coefficient a_j is zero or when the denominatorial gamma function $\Gamma(p+j-Q+1)$ is infinite. We see, then, that a necessary and sufficient condition for $d^Q f_U \neq 0$ is

(5.7.5) $\quad \Gamma(p+j+1-Q) \quad$ is finite for each j for which $a_j \neq 0$.

This awkward condition (5.7.5) may be shown to be equivalent to

(5.7.6) $$f_U - d^{-Q} d^Q f_U = 0;$$

that is, to the condition that the differintegrable unit f_U be regenerated upon the application, first of d^Q, then d^{-Q}. Assuming (5.7.6) temporarily, we find that d^q may then be applied to equation (5.7.4) to give

(5.7.7) $$d^q d^Q f_U = \sum_{j=0}^{\infty} \frac{a_j \Gamma(p+j+1) \Gamma(p+j-Q+1)[x-a]^{p+j-Q-q}}{\Gamma(p+j-Q+1)\Gamma(p+j-Q-q+1)},$$

by another application of equation (5.2.4) valid since $d^Q f_U$ was assumed to be a differintegrable series. With the condition (5.7.6) [or its equivalent (5.7.5)] in effect, we may safely cancel the $\Gamma(p+j-Q+1)$ factors in (5.7.7), arriving at

(5.7.8) $$d^q d^Q f_U = \sum_{j=0}^{\infty} \frac{a_j \Gamma(p+j+1)[x-a]^{p+j-Q-q}}{\Gamma(p+j-Q-q+1)}.$$

On the other hand, the same technique shows that

$$d^{q+Q} f_U = \sum_{j=0}^{\infty} a_j d^{q+Q} f_U = \sum_{j=0}^{\infty} \frac{a_j \Gamma(p+j+1)[x-a]^{p+j-Q-q}}{\Gamma(p+j-Q-q+1)} = d^q d^Q f_U.$$

Thus, the composition rule (5.7.3) is obeyed for the unit f_U as long as condition (5.7.6) is satisfied.[3] However, when (5.7.6) is violated, $d^Q f_U = 0$ so that $d^q d^Q f_U = 0$. On the other hand, it is not necessarily the case that $d^{q+Q} f_U = 0$. For example, we may choose $f_U = x^{-\frac{1}{2}}$, $a = 0$, $Q = \frac{1}{2}$, and $q = -\frac{1}{2}$. Then

$$f_U - d^{-Q} d^Q f_U = x^{-\frac{1}{2}} - d^{-\frac{1}{2}} d^{\frac{1}{2}} x^{-\frac{1}{2}} = x^{-\frac{1}{2}} - d^{-\frac{1}{2}} \frac{\Gamma(\frac{1}{2})}{\Gamma(0)} x^{-1} = x^{-\frac{1}{2}} \neq 0$$

so that condition (5.7.6) is certainly violated. Therefore $d^Q f_U = 0$ and $d^q d^Q f_U = 0$ while $d^{q+Q} f_U = d^0 x^{-\frac{1}{2}} = x^{-\frac{1}{2}} \neq 0$. Generalizing, we easily see the

---

[3] Examination of condition (5.7.5) shows that it, and therefore (5.7.6), are invariably satisfied as long as $Q < 0$ (in fact, as long as $Q < p+1$) since, in that case, $\Gamma(p+j+1-Q)$ is necessarily finite for all j. We see, then, that $d^q d^Q f_U = d^{q+Q} f_U$ at least whenever $Q < 0$, and even when $Q < 1$ if f_U is bounded at the lower limit a.

relationship between $d^q d^Q f_U$ and $d^{q+Q} f_U$ in the case $f_U - d^{-Q} d^Q f_U \neq 0$ to be

(5.7.9) $\qquad 0 = d^q d^Q f_U = d^{q+Q} f_U - d^{q+Q} \{f_U - d^{-Q} d^Q f_U\}.$

The preceding discussion for differintegrable units f_U is summarized in Table 5.7.1.

Table 5.7.1. *Summary of the composition rule[a] for differintegrable units f_U*

	$f_U = 0$	$f_U \neq 0$
$d^Q f_U = 0$	$f_U - d^{-Q} d^Q f_U = 0$ $d^q d^Q f_U = d^{q+Q} f_U = 0$	$f_U - d^{-Q} d^Q f_U \neq 0$ $0 = d^q d^Q f_U = d^{q+Q} f_U$ $- d^{q+Q}[f_U - d^{-Q} d^Q f_U]$
$d^Q f_U \neq 0$	Not attainable	$f_U - d^{-Q} d^Q f_U = 0$ $d^q d^Q f_U = d^{q+Q} f_U.$

[a] The requirements for obedience to this rule are that f_U and $d^Q f_U$ both be differintegrable.

While equation (5.7.9) is a trivial identity for differintegrable units, we shall see that it is less trivial and, therefore, more useful for general differintegrable series. Because equation (5.7.2) is valid for general differintegrable series f if and only if equation (5.7.3) is valid for every differintegrable unit f_U of f, it is straightforward to apply the theory just developed for units f_U to obtain the composition rule for general f. The only difference is that while the conditions

(5.7.10) $\qquad f_U \neq 0 \quad \text{and} \quad f_U - d^{-Q} d^Q f_U = 0$

for units f_U guaranteed that $d^Q f_U \neq 0$, this is no longer the case for arbitrary f. The reason, of course, is that some units of f may satisfy (5.7.10) while others do not. This will make it possible to violate the composition rule (5.7.2) even though

$$f \neq 0 \quad \text{and} \quad d^Q f \neq 0.$$

The condition

(5.7.11) $\qquad f - d^{-Q} d^Q f = 0$

for general differintegrable series f is, however, still necessary and sufficient to guarantee (5.7.2). We mention in passing that for general differintegrable f, as was the case for differintegrable units f_U,

$$d^q d^Q f = d^{q+Q} f,$$

at least when $Q < 0$ (see footnote 3 earlier in this section) and even when $Q < 1$ for functions f bounded at $x = a$. The facts for general differintegrable series f are summarized in Table 5.7.2.

Table 5.7.2. *Summary of the composition rule[a] for* **arbitrary** *differintegrable functions, f*

	$f = 0$	$f \neq 0$
$d^Q f = 0$	$f - d^{-Q} d^Q f = 0$ $d^q d^Q f = d^{q+Q} f = 0$	$f - d^{-Q} d^Q f \neq 0$ $0 = d^q d^Q f = d^{q+Q} f$ $\quad - d^{q+Q}[f - d^{-Q} d^Q f]$
$d^Q f \neq 0$	Not attainable	If $f - d^{-Q} d^Q f = 0$, then $d^q d^Q f = d^{q+Q} f$ If $f - d^{-Q} d^Q f \neq 0$, then $d^q d^Q f = d^{q+Q} f$ $\quad - d^{q+Q}[f - d^{-Q} d^Q f]$

[a] The requirements for obedience to this rule are that f and $d^Q f$ both be differintegrable.

We have noticed previously that, in cases where the composition rule is violated, the equation

(5.7.12) $$d^q d^Q f = d^{q+Q} f - d^{q+Q}\{f - d^{-Q} d^Q f\}$$

relates $d^q d^Q f$ to $d^{q+Q} f$. The utility of equation (5.7.12) as a means of calculating its left-hand side is marginal in the general case. However, there is one important instance in which equation (5.7.12) is very useful: the case when $Q = N$, a positive integer. Indeed, we may then use equations (2.3.12) and (4.4.4) to see that

(5.7.13) $$d^q d^N f = d^{q+N} f - d^{q+N}\{f - d^{-N} d^N f\} = d^{q+N} f - \sum_{k=0}^{N-1} \frac{[x-a]^{k-q-N} f^{(k)}(a)}{\Gamma(k-q-N+1)}.$$

Furthermore, equation (5.7.13) can be established even under the relaxed assumptions[4] that f be N-fold differentiable and that $f^{(k)}(a)$ be finite, $k = 0, 1, \ldots, N-1$. We omit the proof of this assertion, which makes use of the representation (3.6.4). Table 5.7.3 summarizes the composition rule facts for functions f that are N-fold differentiable and whose Nth derivatives are differintegrable.

[4] Since functions such as x^p, $p \leq -1$ are differentiable (with derivative px^{p-1}) but not differintegrable.

86   5 GENERAL PROPERTIES

Table 5.7.3. *Summary of the composition rule[a] for differentiable (but not necessarily differintegrable) functions f*

	$f = 0$	$f \neq 0$
$d^N f = 0$	$f - d^{-N} d^N f = 0$ $d^q d^N f = d^{q+N} f = 0$	$f - d^{-N} d^N f \neq 0$ $0 = d^q d^N f = d^{q+N} f$ $- \sum_{k=0}^{N-1} \dfrac{[x-a]^{k-q-N} f^{(k)}(a)}{\Gamma(k-q-N+1)}$
$d^N f \neq 0$	Not attainable	If $f - d^{-N} d^N f = 0$, then $d^q d^N f = d^{q+N} f$ If $f - d^{-N} d^N f \neq 0$, then $d^q d^N f = d^{q+N} f$ $- \sum_{k=0}^{N-1} \dfrac{[x-a]^{1-q-N} f^{(k)}(a)}{\Gamma(k-q-N+1)}$

[a] The requirements for obedience to this rule are that f be N-fold differentiable and that $d^N f$ be differintegrable.

The general facts just presented about composing d^Q with d^q make it clear that, while the operators d^Q and d^{-Q} are *usually* inverse to each other, this is not always the case. In fact, as we have pointed out repeatedly, one need not look beyond integer orders to find illustrations of this. Indeed, if we choose $f_U = x$, then

$$\frac{d^2 f_U}{dx^2} = 0 \quad \text{and} \quad \frac{d^{-2}}{dx^{-2}} \frac{d^2 f_U}{dx^2} = 0$$

so the the operators d^{-2} and d^2 are certainly not inverse to each other. The difficulty, of course, is that condition (5.7.6) is violated for the differintegrable unit $f_U = x$. Nor is this problem restricted to integer q, Q. In fact, if we choose $Q = \frac{3}{2}$, $q = -\frac{3}{2}$, $a = 0$, and $f_U = x^{\frac{1}{2}}$, then from equation (4.4.4) we see that

$$d^Q f_U = \frac{d^{\frac{3}{2}} x^{\frac{1}{2}}}{dx^{\frac{3}{2}}} = 0 \quad \text{so that} \quad d^q \, d^Q f_U = \frac{d^{-\frac{3}{2}}}{dx^{-\frac{3}{2}}} \frac{d^{\frac{3}{2}} x^{\frac{1}{2}}}{dx^{\frac{3}{2}}} = 0 \quad \text{once again,}$$

while, as we well know,

$$d^{q+Q} f = \frac{d^0 f_U}{dx^0} = f_U = x^{\frac{1}{2}}.$$

This time, too, we have chosen $f_U \neq 0$ but $d^Q f_U = 0$ which guarantees the violation of condition (5.7.6). Finally, to exemplify the possibility of the failure of the composition law even when $f \neq 0$, $d^Q f \neq 0$, we choose $f = \sqrt{x} + 1$,

$N = 1$, $a = 0$, and arbitrary q. This time f is the sum of two differintegrable units and $df = \frac{1}{2}x^{-\frac{1}{2}}$, yet

$$d^q\, df = \frac{\Gamma(\frac{1}{2})}{2\Gamma(-q-\frac{1}{2})} x^{-q-\frac{1}{2}}$$

and

$$d^{q+1}f = \frac{\Gamma(\frac{3}{2})}{\Gamma(-q-\frac{1}{2})} x^{-q-\frac{1}{2}} + \frac{1}{\Gamma(-q)} x^{-q-1}$$

$$= \frac{\Gamma(\frac{1}{2})}{2\Gamma(-q-\frac{1}{2})} x^{-q-\frac{1}{2}} + \frac{1}{\Gamma(-q)} x^{-q-1}.$$

As must be so, the condition (5.7.11) is violated ($f - d^{-1}\,df = 1$) and equation (5.7.13) correctly relates $d^q\,df$ and $d^{q+1}f$.

The discussion just completed also reveals the dangers that lurk when one inquires about the commutativity of differintegral operators ($d^q\,d^Q = d^Q\,d^q$) or the invertibility of differintegration ($d^Q f = g$ implies $f = d^{-Q}g$). Of course the latter comes into play in attempting to solve differintegral equations of arbitrary order; we shall have more to say about this subject in Chapter 9. The results of the present section may be used to derive conditions under which commutativity holds, provided, as always, that the operators are applied to a suitably restricted class of functions.

5.8 DEPENDENCE ON LOWER LIMIT

We have postponed until now any discussion of the manner in which $d^q f/[d(x-a)]^q$ depends on the lower limit a. In the present section we derive a formula that exhibits this dependence in a rather concise way, at least for analytic functions ϕ.

Assume $a < b < x$ and that $\phi(y)$ is any function analytic in the interval $a \leq y \leq x$. Writing

(5.8.1) $$\Delta = \frac{d^q \phi}{[d(x-a)]^q} - \frac{d^q \phi}{[d(x-b)]^q},$$

one might expect that the difference Δ is expressible in terms of integrals of $\phi(y)$ confined to the interval $a \leq y \leq b$. The correct formula is

(5.8.2) $$\Delta = \sum_{l=1}^{\infty} \frac{d^{q+l}[1]}{[d(x-b)]^{q+l}} \frac{d^{-l}\phi(b)}{[d(b-a)]^{-l}}.$$

88   **5 GENERAL PROPERTIES**

To prove this we first establish equation (5.8.2) for $q < 0$ making use of (3.6.2), and then utilize the identity theorem for analytic functions to argue that the result is valid in general.

Thus, assuming $q < 0$, we write

(5.8.3)
$$\begin{aligned}
\Delta &= \frac{1}{\Gamma(-q)} \int_a^x \frac{\phi(y)\,dy}{[x-y]^{q+1}} - \frac{1}{\Gamma(-q)} \int_b^x \frac{\phi(y)\,dy}{[x-y]^{q+1}} \\
&= \frac{1}{\Gamma(-q)} \int_a^b \frac{\phi(y)\,dy}{[x-y]^{q+1}} \\
&= \frac{1}{\Gamma(-q)} \int_a^b \frac{\phi(y)\,dy}{[x-b+b-y]^{q+1}} \\
&= \frac{1}{\Gamma(-q)} \int_a^b \left[\sum_{l=0}^{\infty} \binom{-1-q}{l} [x-b]^{-1-q-l} [b-y]^l \right] \phi(y)\,dy
\end{aligned}$$

upon binomial expansion of $[x-b+b-y]^{-1-q}$. When j is replaced by l and q by $q+l$, equation (1.3.16) becomes

(5.8.4)
$$\binom{-1-q}{l} = \frac{\Gamma(-q)}{\Gamma(-q-l)\Gamma(l+1)}.$$

Putting (5.8.4) into (5.8.3) produces

$$\begin{aligned}
\Delta &= \int_a^b \sum_{l=0}^{\infty} \frac{[x-b]^{-1-q-l}}{\Gamma(-q-l)} \frac{[b-y]^l}{\Gamma(l+1)} \phi(y)\,dy \\
&= \sum_{l=0}^{\infty} \frac{d^{q+l+1}[1]}{[d(x-b)]^{q+l+1}} \int_a^b \frac{\phi(y)\,dy}{\Gamma(l+1)[b-y]^{-l}} \\
&= \sum_{l=0}^{\infty} \frac{d^{q+l+1}[1]}{[d(x-b)]^{q+l+1}} \frac{d^{-l-1}\phi(b)}{[d(b-a)]^{-l-1}} = \sum_{l=0}^{\infty} \frac{d^{q+l}[1]}{[d(x-b)]^{q+l}} \frac{d^{-l}\phi(b)}{[d(b-a)]^{-l}},
\end{aligned}$$

establishing (5.8.2) for $q < 0$. We now argue that, since both sides of equation (5.8.2) are analytic in q, the equation is valid for all q by the identity theorem for analytic functions (see the discussion in Section 3.2).

Some special cases of formula (5.8.2) require consideration. When q is zero or a positive integer, all the derivatives of unity in the formula vanish, so that

$$\Delta = 0, \quad q = 0, 1, 2, \ldots.$$

When $q = -1$, all the derivatives vanish except for the first, whence

$$\Delta = \frac{d^{-1}\phi(b)}{[d(b-a)]^{-1}} \int_a^b \phi(y)\,dy, \quad q = -1.$$

For all other values of q, Δ is nonzero and its value depends not only upon a and b, but also on x. For example, when q is a negative integer $-n$, we find

$$\Delta = \sum_{l=1}^{n} \frac{d^{l-n}[1]}{[d(x-b)]^{l-n}} \frac{d^{-l}\phi(b)}{[d(b-z)]^{-l}}$$

$$= \sum_{l=1}^{n} \frac{[x-b]^{n-l}}{\Gamma(1-l+n)} \frac{d^{-l}\phi(b)}{[d(b-a)]^{-l}}, \qquad q - n = -1, -2, -3, \ldots.$$

This result is identical with that derived in Section 2.4 by a classical argument.

5.9 TRANSLATION

By the translation of a function f we mean the replacement of $f(x)$ by the function $f(A + x)$, where A is a constant which we take to be positive. In this section we seek a rule for evaluating the effect of a differintegral operator on a translated function, that is, we seek to relate

$$\frac{d^q f(A+x)}{[d(x-a)]^q} \qquad \text{to} \qquad \frac{d^q f(x)}{[d(x-a)]^q},$$

assuming, of course, that f is defined wherever needed to have these differintegrals make sense, i.e., between $\min(a, a+A)$ and $\max(x, x+A)$.

As usual, the Riemann–Liouville definition is the most tractable and from it we see immediately that

$$\frac{d^q f(x+A)}{[d(x-a)]^q} = \frac{1}{\Gamma(-q)} \int_a^x \frac{f(y+A)\, dy}{[x-y]^{q+1}} = \frac{1}{\Gamma(-q)} \int_{a+A}^{x+A} \frac{f(Y)\, dY}{[x+A-y]^{q+1}},$$

where $Y = y + A$. It is evident that translation by a distance A is equivalent to a shift in the upper limit from x to $x + A$ and a shift in the lower limit from a to $a + A$. Representing the effect of the latter shift by Δ, we therefore find

(5.9.1)

$$\frac{d^q f(x+A)}{[d(x-a)]^q} = \frac{d^q f(x+A)}{[d(x+A-a)]^q} - \Delta$$

$$= \frac{d^q f(x+A)}{[d(x+A-a)]^q} - \sum_{l=1}^{\infty} \frac{d^{q+l}[1]}{[d(x+A-a)]^{q+l}} \frac{d^{-l}f(a+A)}{[d(a+A-a)]^{-l}},$$

where the results of Section 5.8 have been used to evaluate Δ. Though the use of the Riemann–Liouville definition requires $q < 0$, the usual analyticity argument based on the identity theorem serves to remove this restriction.

Formula (5.9.1) involves an infinite sum that is not, in general, amenable to expression in closed form. Translation, then, represents a process which is difficult to handle in our generalized calculus. Fortunately, however, the properties of many of the important functions considered in Chapter 6 enable the difficulties inherent in (5.9.1) to be short-circuited.

5.10 BEHAVIOR NEAR LOWER LIMIT

When q is a positive integer or zero, the operator $d^q/[d(x - a)]^q$ is, as we have seen, local, i.e.,

$$\frac{d^q f}{[d(x - a)]^q} = \frac{d^q f}{dx^q}, \qquad q = n = 0, 1, 2, \ldots,$$

and the behavior of $d^q f/[d(x - a)]^q$ near $x = a$ is unexceptional. For all other values of q, however, it will now be demonstrated that $d^q f/[d(x - a)]^q$ usually approaches either zero or infinity as x approaches a, for all differintegrable series f. If f is a differintegrable series it can be decomposed as a finite sum of differintegrable series units f_U,

$$f_U = [x - a]^p \sum_{j=0}^{\infty} a_j [x - a]^j, \qquad p > -1, \quad a_0 \neq 0.$$

We shall demonstrate that $d^q f_U/[d(x - a)]^q$ normally approaches either zero or infinity as x approaches a for all such units f_U. Thus, the same conclusion will be valid for f by virtue of the linearity of $d^q/[d(x - a)]^q$.

If f_U is a differintegrable series unit, we have by equation (5.2.4),

(5.10.1) $$\frac{d^q f_U}{[d(x - a)]^q} = \sum_{j=0}^{\infty} \frac{a_j \Gamma(p + j + 1)}{\Gamma(p + j - q + 1)} [x - a]^{p+j-q}.$$

We know that $1/\Gamma(p + j - q + 1)$ is always finite, while $\Gamma(p + j + 1)$ is finite since $p > -1$. Thus, the right-hand side of (5.10.1) is dominated by its first term for small $x - a$. That is

(5.10.2)
$$\lim_{x \to a} \left\{ \frac{d^q f_U}{[d(x - a)]^q} \right\} = \lim_{x \to a} \left\{ \frac{[x - a]^{p-q} a_0 \Gamma(p + 1)}{\Gamma(p - q + 1)} \right\} = \begin{cases} 0, & p - q > 0, \\ a_0 \Gamma(p + 1), & p - q = 0, \\ \infty, & p - q < 0, \end{cases}$$

since $a_0 \neq 0$.

5.11 BEHAVIOR FAR FROM LOWER LIMIT

In the present section we restrict attention to analytic functions ϕ. If $x \gg a$, we may write

$$[x - a]^{k-q} = x^{k-q}\left[1 - \frac{a}{x}\right]^{k-q}$$
$$= x^{k-q}\left[1 - \frac{[k-q]a}{x} + O\left(\frac{a^2}{x^2}\right)\right]$$
$$\sim x^{k-q} + \frac{[q-k]ax^k}{x^{q+1}}$$

and substitution into equation (3.5.3) yields [after use of (1.3.8)]

(5.11.1)
$$\frac{d^q\phi}{[d(x-a)]^q} \approx \frac{\Gamma(q+1)\sin(\pi q)}{\pi}\left[\sum_{k=0}^{\infty}\frac{[-]^k x^{k-q}\phi^{(k)}}{[q-k]k!} + \frac{a}{x^{q+1}}\sum_{k=0}^{\infty}\frac{[-]^k x^k \phi^{(k)}}{k!}\right]$$
$$= \frac{d^q\phi}{dx^q} + \frac{a\Gamma(q+1)\sin(\pi q)\phi(0)}{\pi x^{q+1}}.$$

We observe that if q is a positive integer, the final term in equation (5.11.1) vanishes, reminding us once again of the local character of integer order derivatives.

CHAPTER 6

DIFFERINTEGRATION OF MORE COMPLEX FUNCTIONS

Chapter 4 was devoted to the differintegration of certain simple functions. Having now developed, in Chapter 5, some general rules governing differintegration, we are in a position to tackle more difficult functions. The most powerful weapon in our armory is the rule

$$(6.0.1) \quad \frac{d^q}{[d(x-a)]^q} \sum_{j=0}^{\infty} a_j [x-a]^{p+[j/n]}$$

$$= \sum_{j=0}^{\infty} a_j \frac{\Gamma\left(\frac{pn+j+n}{n}\right)}{\Gamma\left(\frac{pn-qn+j+n}{n}\right)} [x-a]^{p-q+[j/n]}$$

for $p > -1$, established in Section 5.2 for the differintegral of the function $\sum a_j[x-a]^{p+[j/n]}$. As will be recalled from the discussion in Section 3.1, such a function belongs to the class we have termed "differintegrable series" provided n is a positive integer, p exceeds -1, and a_0 is nonzero.

In early sections of this chapter we shall differintegrate using an arbitrary lower limit a; but, to avoid the added complexity introduced by this generality, some later sections will adopt a lower limit of zero.

6.1 THE BINOMIAL FUNCTION $[C-cx]^p$

The binomial theorem permits, upon writing $C - cx = C - ac - c[x-a]$, the power-series expansion

$$[C-cx]^p = \sum_{j=0}^{\infty} \frac{\Gamma(p+1)}{\Gamma(j+1)\Gamma(p-j+1)} [-c]^j [C-ac]^{p-j} [x-a]^j$$

94   6 DIFFERINTEGRATION OF MORE COMPLEX FUNCTIONS

in $x - a$, provided that the quantity X, defined as $c[x - a]/[C - ca]$, lies in the range $-1 < X < +1$. Note that this expansion is valid even if p is a positive integer, but that in this event the sum is automatically finite, all terms for which j exceeds p having infinite denominators. Equation (6.0.1) may be applied straightforwardly to this sum, leading to the result

$$\frac{d^q[C - cx]^p}{[d(x - a)]^q} = \frac{[C - ca]^p[x - a]^{-q}}{\Gamma(-p)} \sum_{j=0}^{\infty} \frac{\Gamma(j - p)}{\Gamma(j - q + 1)} \left[\frac{c[x - a]}{C - ca}\right]^j$$

after use of the $\Gamma(p + 1)/\Gamma(p - j + 1) = [-]^j \Gamma(j - p)/\Gamma(-p)$ identity and removal of all j-independent factors from within the summation. The most concise representation of the summed terms is as an incomplete beta function (see Section 1.3) of argument X, yielding

$$\frac{d^q[C - cx]^p}{[d(x - a)]^q} = \frac{c^q[C - cx]^{p-q}}{\Gamma(-q)} B_X(-q, q - p)$$

as the final result.

For the case in which $a = 0$, and $C = c = 1$, the simple result

(6.1.1) $$\frac{d^q[1 - x]^p}{dx^q} = \frac{[1 - x]^{p-q}}{\Gamma(-q)} B_x(-q, q - p)$$

emerges.

6.2 THE EXPONENTIAL FUNCTION $\exp(C - cx)$

With C and c as arbitrary constants, the power-series expansion

$$\exp(C - cx) = \exp(C - ca) \sum_{j=0}^{\infty} \frac{\{-c[x - a]\}^j}{\Gamma(j + 1)}$$

is valid for all $x - a$. Differintegration term by term with respect to $c[x - a]$ yields

$$\frac{d^q \exp(C - cx)}{[d(cx - ca)]^q} = \{c[x - a]\}^{-q} \exp(C - ca) \sum_{j=0}^{\infty} \frac{\{-c[x - a]\}^j}{\Gamma(j - q + 1)}.$$

The sum may be expressed as an incomplete gamma function [see equation (1.3.26)] of argument $-c[x - a]$ and parameter $-q$. The final result appears as

(6.2.1) $$\frac{d^q \exp(C - cx)}{[d(x - a)]^q} = \frac{\exp(C - cx)}{[x - a]^q} \gamma^*(-q, -c[x - a])$$

after use of the scale change relationship (5.4.2) to replace the variable of differintegration by $x - a$.

Since $\gamma^*(-n, y) = y^n$ for nonnegative integer n, the above result is seen to reduce to the well-known formula for multiple differentiation of an exponential function. Reduction to the simple formula

(6.2.2) $$\frac{d^q \exp(\pm x)}{dx^q} = \frac{\exp(\pm x)}{x^q} \gamma^*(-q, \pm x)$$

occurs on substituting $C = a = 0$ and $c = \mp 1$ into the general result.

The incomplete gamma function has an asymptotic expansion which permits us to write

$$\frac{\gamma^*(-q, y)}{y^q} \sim 1 - \frac{\exp(-y)}{\Gamma(-q)y^{q+1}} \left[1 - \frac{q+1}{y} + O(y^{-2}) \right]$$

for large y. This expansion may be invoked to evaluate the $a \to -\infty$ limit of equation (6.2.1). Thus, if we choose $C = 0$, $c = -v$,

$$\frac{d^q \exp(vx)}{[d(x + \infty)]^q} = v^q \exp(vx).$$

This simple result for differintegration with a lower limit of $-\infty$ provided the basis for much of the work of Liouville and Weyl, about which we wrote in Section 1.1. Functions other than exponentials, however, seldom yield finite differintegrals when the lower limit is minus infinity and we shall have no occasion again to make use of these "Weyl differintegrals."

The functions considered in this section and the last, $[C - cx]^p$ and $\exp(C - cx)$, are both analytic in $x - a$. The functions treated in the next section, while they are not necessarily analytic, are differintegrable and so are still subject to the term-by-term differintegration rule (6.0.1).

6.3 THE FUNCTIONS $x^q/[1-x]$ AND $x^p/[1-x]$ AND $[1-x]^{q-1}$

By use of the binomial expansion of $[1 - x]^{-1}$ and the technique of term-by-term differintegration, we arrive at

$$\frac{d^q}{dx^q} \left[\frac{x^q}{1 - x} \right] = \sum_{j=0}^{\infty} \frac{d^q}{dx^q} x^{j+q}$$

as a formula expressing the effect of the d^q/dx^q operator with the lower limit zero on the $x^q/[1 - x]$ function, subject to the proviso that x not exceed unity in magnitude. Provided also that q exceeds -1, the rules of Section 4.4 permit differintegration of the powers of x and lead to

$$\frac{d^q}{dx^q} \left[\frac{x^q}{1-x} \right] = \sum_{j=0}^{\infty} \frac{\Gamma(j+q+1)}{\Gamma(j+1)} x^j = \Gamma(q+1) \sum_{j=0}^{\infty} \binom{-q-1}{j} [-x]^j.$$

96 6 DIFFERINTEGRATION OF MORE COMPLEX FUNCTIONS

Identification of the sum as a binomial expansion produces

(6.3.1) $$\frac{d^q}{dx^q}\left[\frac{x^q}{1-x}\right] = \frac{\Gamma(q+1)}{[1-x]^{q+1}}$$

as the simple final result.

The technique for differintegrating $x^p/[1-x]$ follows such a similar course that it will suffice to cite one intermediate and the final result

(6.3.2) $$\frac{d^q}{dx^q}\left[\frac{x^p}{1-x}\right] = x^{p-q}\sum_{j=0}^{\infty}\frac{\Gamma(j+p+1)x^j}{\Gamma(j+p-q+1)} = \frac{\Gamma(p+1)B_x(p-q,q+1)}{\Gamma(p-q)[1-x]^{q+1}},$$

together with the restrictions, namely, $0 < x < 1$ and $p > -1$, which were assumed during the derivation.

As an illustration of the utility of the composition rule (Section 5.7) in finding differintegrals, consider the effect of the d^{-q}/dx^{-q} operator applied to each member of equation (6.3.1). The composition rule may readily be applied when $q < 0$ to yield

$$\frac{x^q}{1-x} = \Gamma(q+1)\frac{d^{-q}}{dx^{-q}}[1-x]^{-q-1}.$$

Extension to $q < 1$ follows[1] by the boundedness at zero of the function $x^q/[1-x]$. [Recall also that $q > -1$ was assumed in the derivation of (6.3.1).] Rearrangement and reversal of the sign of q then produces

$$\frac{d^q}{dx^q}[1-x]^{q-1} = \frac{x^{-q}}{\Gamma(1-q)[1-x]}$$

with $|q| < 1$. This result may be regarded as a special case of equation (6.1.1), from which it may be derived alternatively.

6.4 THE HYPERBOLIC AND TRIGONOMETRIC FUNCTIONS $\sinh(\sqrt{x})$ AND $\sin(\sqrt{x})$

The differintegrals of the hyperbolic and circular sines of the square root of x are particularly interesting examples of the differintegration of nonanalytic functions. They provide a foretaste of the capability of differintegration to

---

[1] See Chapter 5, footnote 3.

interrelate important transcendental functions, a subject which is considered in detail in Chapter 9.

The now familiar technique of series expansion followed by term-by-term differintegration gives

$$\frac{d^q}{dx^q}\sinh(\sqrt{x}) = \sum_{j=0}^{\infty} \frac{d^q}{dx^q}\left[\frac{x^{j+\frac{1}{2}}}{\Gamma(2j+2)}\right] = \sum_{j=0}^{\infty} \frac{\Gamma(j+\frac{3}{2})x^{j-q+\frac{1}{2}}}{\Gamma(2j+2)\Gamma(j-q+\frac{3}{2})}.$$

The simplification

$$\frac{\Gamma(j+\frac{3}{2})}{\Gamma(2j+2)} = \frac{\sqrt{\pi}}{2^{2j+1}\Gamma(j+1)}$$

is a consequence of the duplication property of gamma functions and permits us to write

(6.4.1) $$\frac{d^q}{dx^q}\sinh(\sqrt{x}) = \frac{\sqrt{\pi}\,x^{\frac{1}{2}-q}}{2}\sum_{j=0}^{\infty}\frac{[\frac{1}{4}x]^j}{\Gamma(j-q+\frac{3}{2})\Gamma(j+1)}$$
$$= \tfrac{1}{2}\sqrt{\pi}\,[2\sqrt{x}]^{\frac{1}{2}-q}I_{\frac{1}{2}-q}(\sqrt{x}),$$

where $I_{\frac{1}{2}-q}(\sqrt{x})$ denotes the $(\frac{1}{2}-q)$th-order hyperbolic Bessel function of argument $\sqrt{x}$.

The differintegral of the circular sine is derived in a strictly analogous way, being

(6.4.2) $$\frac{d^q}{dx^q}\sin(\sqrt{x}) = \tfrac{1}{2}\sqrt{\pi}\,[2\sqrt{x}]^{\frac{1}{2}-q}J_{\frac{1}{2}-q}(\sqrt{x})$$

and involving an ordinary Bessel function of the first kind. These last two formulas constitute generalizations of Rayleigh's formulas (Abramowitz and Stegun, 1964, pp. 439, 445).

6.5 THE BESSEL FUNCTIONS

This section will evaluate the differintegrals of the functions

$$x^{v/2}J_v(2\sqrt{x}) \quad \text{and} \quad x^{v/2}I_v(2\sqrt{x}),$$

where $J_v(\ )$ and $I_v(\ )$ are the vth-order Bessel and modified Bessel functions. This exercise will exemplify the fact that some formulas of the classical calculus generalize unchanged into the fractional calculus, whereas others do not.

6 DIFFERINTEGRATION OF MORE COMPLEX FUNCTIONS

The vth-order Bessel function is defined via the series

(6.5.1) $$J_v(2\sqrt{x}) \equiv x^{v/2} \sum_{j=0}^{\infty} \frac{[-x]^j}{\Gamma(j+1)\Gamma(j+v+1)}.$$

Therefore,

$$x^{v/2} J_v(2\sqrt{x}) = \sum_{j=0}^{\infty} \frac{[-]^j x^{j+v}}{\Gamma(j+1)\Gamma(j+v+1)}$$

and is seen to be a differintegrable function provided $v > -1$. Performing the differintegration with a lower limit of zero yields

(6.5.2) $$\frac{d^q}{dx^q}\{x^{v/2} J_v(2\sqrt{x})\} = \sum_{j=0}^{\infty} \frac{[-]^j x^{j+v-q}}{\Gamma(j+1)\Gamma(j+v-q+1)}$$
$$= x^{[v-q]/2} J_{v-q}(2\sqrt{x}),$$

a simple and appealing result.

The proof that

(6.5.3) $$\frac{d^q}{dx^q}\{x^{v/2} I_v(2\sqrt{x})\} = x^{[v-q]/2} I_{v-q}(2\sqrt{x})$$

follows an identical pattern since the definition of the modified Bessel function mirrors equation (6.5.1) except that the alternating signs within the summand are missing.

If we set $z \equiv 2\sqrt{x}$ in (6.5.2), transformation to

(6.5.4) $$\frac{d^q}{[z\,dz]^q}\{z^v J_v(z)\} = z^{v-q} J_{v-q}(z)$$

occurs. This result has long been known in the classical calculus [see Abramowitz and Stegun, (1964, p. 361)], where it is restricted, of course, to integer q. It is seen to generalize unchanged into the fractional calculus.

A classical result, complementary to (6.5.4), is

(6.5.5) $$\frac{d^q}{[z\,dz]^q}\left\{\frac{J_v(z)}{z^v}\right\} = [-]^q \frac{J_{v+q}(z)}{z^{v+q}}$$

for integer q. This result clearly cannot generalize unchanged into the fractional calculus, for whereas the left-hand side is real (for real z) the right-hand side would necessarily be complex for many q values. The appropriate extension of equation (6.5.5) for the special values $q = \pm\frac{1}{2}$ is given in Section 7.7.

6.6 HYPERGEOMETRIC FUNCTIONS

A definition of a generalized hypergeometric function was given in Section 2.10. Formulas were there developed showing the effect of integer order differintegration on a generalized hypergeometric function and on the product of such a function with x^p. The technique of Section 5.2 shows that these formulas [(2.10.2) and (2.10.3)] generalize unchanged if q is unrestricted. That is,

(6.6.1) $$\frac{d^q}{dx^q}\left[x \frac{b_1, b_2, \ldots, b_K}{c_1, c_2, \ldots, c_L}\right] = x^{-q}\left[x \frac{0, b_1, b_2, \ldots, b_K}{-q, c_1, c_2, \ldots, c_L}\right]$$

and

(6.6.2) $$\frac{d^q}{dx^q}\left\{x^p\left[x \frac{b_1, b_2, \ldots, b_K}{c_1, c_2, \ldots, c_L}\right]\right\}$$
$$= x^{p-q}\left[x \frac{p, b_1, b_2, \ldots, b_K}{p-q, c_1, c_2, \ldots, c_L}\right], \quad p > -1$$

for all q.

It will be recognized that all hypergeometric functions of argument x (or of argument $x^{m/n}$, where m and n are integers) fall into our class of differintegrable series, as defined in Section 3.1. This class also embraces all products of a hypergeometric function with x^p (for $p > -1$) or with a second hypergeometric function.

Inasmuch as

$$[1-x]^p = \frac{1}{\Gamma(-p)}\left[x \frac{-1-p}{0}\right]$$

and

$$\frac{[1-x]^{p-q}}{\Gamma(-q)}\mathbf{B}_x(-q, q-p) = \frac{x^{-q}}{\Gamma(-p)}\left[x \frac{-1-p}{-q}\right],$$

result (6.1.1) is seen to be nothing but a special case of formula (6.6.1) for differintegration of a generalized hypergeometric function, coupled with the rule for canceling equal numeratorial and denominatorial parameters. Result (6.2.2) is an even simpler special case of the same formula.

Likewise, all other results we have thus far derived in this chapter [formulas (6.3.1), (6.3.2), (6.4.1), (6.4.2), (6.5.2), and (6.5.3)] may easily be shown to be special instances of the formula (6.6.2) for differintegration of a product of a power function x^p (with $p > -1$) and a generalized hypergeometric

100   6 DIFFERINTEGRATION OF MORE COMPLEX FUNCTIONS

function. Consider formula (6.4.1) as an example. Using the duplication formula (1.3.9), the hyperbolic sine of $\sqrt{x}$ may be written as a $\frac{0}{2}$ hypergeometric function,

$$\sinh(\sqrt{x}) = \frac{\sqrt{\pi x}}{2} \left[\tfrac{1}{4} x \, \frac{}{0, \tfrac{1}{2}} \right],$$

so that, utilizing the scale-change theorem, we find

$$\frac{d^q}{dx^q} \sinh(\sqrt{x}) = \frac{\sqrt{\pi}}{4^q} \frac{d^q}{[d(x/4)]^q} \left\{ \sqrt{\frac{x}{4}} \left[\tfrac{1}{4} x \, \frac{}{0, \tfrac{1}{2}} \right] \right\}$$

$$= \frac{\sqrt{\pi}}{4^q} \left[\frac{x}{4} \right]^{\tfrac{1}{2}-q} \left[\tfrac{1}{4} x \, \frac{\tfrac{1}{2}}{\tfrac{1}{2}-q, 0, \tfrac{1}{2}} \right]$$

$$= \frac{\sqrt{\pi}}{2} x^{\tfrac{1}{2}-q} \left[\tfrac{1}{4} x \, \frac{}{\tfrac{1}{2}-q, 0} \right] = \tfrac{1}{2} \sqrt{\pi} \, [2\sqrt{x}]^{\tfrac{1}{2}-q} I_{\tfrac{1}{2}-q}(\sqrt{x})$$

because of the expressibility of a modified Bessel function as the $\frac{0}{2}$ hypergeometric function,

$$I_\nu(\sqrt{x}) = \left[\frac{x}{4} \right]^{\nu/2} \left[\tfrac{1}{4} x \, \frac{}{0, \nu} \right].$$

Use was made of the scale-change property (Section 5.4) in the preceding paragraph. The procedure may be generalized,

$$(6.6.3) \quad \frac{d^q}{dx^q} \left\{ x^p \left[\beta x \, \frac{b_1, b_2, \ldots, b_K}{c_1, c_2, \ldots, c_L} \right] \right\}$$

$$= \beta^{q-p} \frac{d^q}{[d(\beta x)]^q} \left\{ [\beta x]^p \left[\beta x \, \frac{b_1, b_2, \ldots, b_K}{c_1, c_2, \ldots, c_L} \right] \right\}$$

$$= x^{p-q} \left[\beta x \, \frac{p, b_1, b_2, \ldots, b_K}{p-q, c_1, c_2, \ldots, c_L} \right], \quad p > -1,$$

to show that the differintegration properties of a hypergeometric of argument equal to a constant multiplied by x are not affected by the magnitude of the constant.

Thus far we. have treated the generalized hypergeometric functions of argument equal to x, the variable of differintegration, or some constant multiple thereof. We now turn to situations in which the argument is a root $x^{1/n}$ or power x^n of x, n being a (typically small) positive integer.

6.6 HYPERGEOMETRIC FUNCTIONS 101

We demonstrated in Section 2.10 that a $\frac{K}{L}$ hypergeometric function of argument $x^{1/n}$ could be equated to a sum of n hypergeometrics, each generally having the complexity $\frac{nK}{nL}$ and argument x. The problem of differintegrating the original function is thus solved. Formula (2.10.4) was the relevant equation and we use it in the following development of the differintegration properties of the function $[1 - x^{\frac{1}{3}}]^{-1}$:

$$\frac{d^q}{dx^q}\left\{\frac{x^p}{1-x^{\frac{1}{3}}}\right\} = \frac{d^q}{dx^q}\left\{x^p\left[x^{\frac{1}{3}} \underline{\quad}\right]\right\}$$

$$= \frac{d^q}{dx^q}\left\{x^p\left[x \underline{\quad}\right]\right\} + \frac{d^q}{dx^q}\left\{x^{p+\frac{1}{3}}\left[x \underline{\quad}\right]\right\} + \frac{d^q}{dx^q}\left\{x^{p+\frac{2}{3}}\left[x \underline{\quad}\right]\right\}$$

$$= x^{p-q}\left[x \frac{p}{p-q}\right] + x^{p-q+\frac{1}{3}}\left[x \frac{p+\frac{1}{3}}{p-q+\frac{1}{3}}\right]$$

$$+ x^{p-q+\frac{2}{3}}\left[x \frac{p+\frac{2}{3}}{p-q+\frac{2}{3}}\right].$$

Finally, we analyze the differintegration properties of the generalized hypergeometric function of argument x^n, using the example of a function of $\frac{1}{1}$ complexity, namely, $\left[x^n \frac{b}{c}\right]$. First, however, we shall state the result,

$$(6.6.4) \qquad \frac{\Gamma(jn + p + 1)}{\Gamma(jn + p - q + 1)} = n^q \prod_{i=1}^{n} \frac{\Gamma\left(j + \frac{p+i}{n}\right)}{\Gamma\left(j + \frac{p-q+1}{n}\right)},$$

of applying the Gauss multiplication formula (1.3.10) to the left-hand member of (6.6.1). We are now in a position to develop the following proof:

$$\frac{d^q}{dx^q} x^p \left[x^n \frac{b}{c}\right]$$

$$= \sum_{j=0}^{\infty} \frac{d^q}{dx^q} x^{jn+p} \frac{\Gamma(j+b+1)}{\Gamma(j+c+1)}$$

$$= x^{p-q} \sum_{j=0}^{\infty} x^{jn} \frac{\Gamma(jn+p+1)\Gamma(j+b+1)}{\Gamma(jn+p-q+1)\Gamma(j+c+1)}$$

$$= n^q x^{p-q}\left[x^n \frac{\frac{p-n+1}{n}, \frac{p-n+2}{n}, \ldots, \frac{p-1}{n}, \frac{p}{n}, b}{\frac{p-q-n+1}{n}, \frac{p-q-n+2}{n}, \ldots, \frac{p-q}{n}, c}\right].$$

6 DIFFERINTEGRATION OF MORE COMPLEX FUNCTIONS

Extension to a hypergeometric function of $\frac{K}{L}$ complexity is obvious, and is seen to yield a differintegral of $\frac{K+n}{L+n}$ complexity in the absence of parametric cancellation. For the important $n = 2$ case, the general result reads

$$(6.6.5) \qquad \frac{d^q}{dx^q}\left\{x^p\left[x^2\,\frac{b_1, b_2, \ldots, b_K}{c_1, c_2, \ldots, c_L}\right]\right\}$$

$$= 2^q x^{p-q}\left[x^2\,\frac{\dfrac{p-1}{2},\dfrac{p}{2},b_1,b_2,b_3,\ldots,b_K}{\dfrac{p-q-1}{2},\dfrac{p-q}{2},c_1,\ldots,c_L}\right].$$

6.7 LOGARITHMS

Its recognition as the hypergeometric function

$$\ln(x+1) = x\left[-x\,\frac{0}{1}\right]$$

permits a ready differintegration of the logarithm function of argument $x + 1$. Thus

$$\frac{d^q \ln(x+1)}{dx^q} = x^{1-q}\left[-x\,\frac{0}{1-q}\right]$$

on application of rule (6.6.2) and cancellation of a unity parameter. Surprisingly, since it occurs ubiquitously in the generalized calculus, the simple hypergeometric function

$$\left[x\,\frac{0}{c}\right]$$

lacks a generic name, though instances of it are widespread (see Chapter 9).

Because of the simplicity of the chain rule for integer-order derivatives (Section 2.6) a change of function variable is readily accomplished in the classical differential calculus. As we saw in Section 5.6, however, no simple analog of the chain rule exists in the general calculus, a fact which prevents evaluation of

$$\frac{d^q \ln(x)}{dx^q} \qquad \text{by way of} \qquad \frac{d^q \ln(x+1)}{dx^q}.$$

Likewise, if we try to accomplish the same derivation by noting the equivalence of

$$\frac{d^q \ln(x)}{dx^q} \quad \text{to} \quad \frac{d^q \ln(X+1)}{[d(X+1)]^q},$$

(with $X = x - 1$) and attempt to relate the latter to

$$\frac{d^q \ln(X+1)}{dX^q},$$

we are impeded by the complexity of the rule (Section 5.8) for shifting the lower limit of differintegration. In fact, in the generalized calculus, differintegrals of $\ln(x)$ and $\ln(x+1)$ are astonishingly unrelated. We shall find this to be quite a common state of affairs—for example such apparently similar functions as $\sin(x)$ and $\sin(\sqrt{x})$ behave quite differently on general differintegration—and we shall learn not to be surprised thereby.

To differintegrate $\ln(x)$ we start with the Riemann–Liouville formulation and then apply the $v = [x - y]/x$ variable change:

$$\frac{d^q \ln(x)}{dx^q} = \frac{1}{\Gamma(-q)} \int_0^x \frac{\ln(y)\, dy}{[x-y]^{q+1}}, \quad q < 0$$

$$= \frac{x^{-q} \ln(x)}{\Gamma(-q)} \int_0^1 \frac{dv}{v^{q+1}} + \frac{x^{-q}}{\Gamma(-q)} \int_0^1 \frac{\ln(1-v)\, dv}{v^{q+1}},$$

to produce two definite integrals. The first evaluates trivially to $1/[-q]$, while the second yields to the parts-integration,

$$\int_0^1 \frac{\ln(1-v)\, dv}{v^{q+1}} = \frac{1}{q} \int_0^1 \ln(1-v)\, d(1 - v^{-q})$$

$$= \frac{[1 - v^{-q}] \ln(1-v)}{q} \bigg|_0^1 - \frac{1}{q} \int_0^1 \frac{1 - v^{-q}}{1 - v}\, dv$$

$$= 0 - \frac{\gamma + \psi(1-q)}{q},$$

where we have made use of equations (1.3.35) and (1.3.36) in deriving the term containing the psi function of $1 - q$. Putting these results together, and employing the recurrence (1.3.2),

(6.7.1) $$\frac{d^q \ln(x)}{dx^q} = \frac{x^{-q}}{\Gamma(1-q)} [\ln(x) - \gamma - \psi(1-q)]$$

is obtained. Though derived for $q < 0$, the usual appeal to analyticity establishes result (6.7.1) for all q, though the expression is indeterminate as formulated for q a positive integer.[2] Note the reduction of (6.7.1) to the classical result

$$\frac{d^{-n}\ln(x)}{dx^{-n}} = \frac{x^n}{n!}\left[\ln(x) - \sum_{j=1}^{n}\frac{1}{j}\right]$$

when q is a negative integer, in consequence of equations (1.3.34) and (1.3.35). Figure 6.7.1 shows some differintegrals of $\ln(x)$.

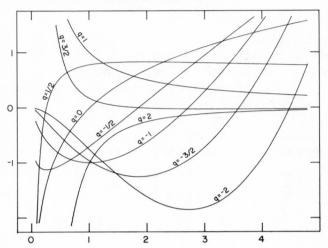

FIG. 6.7.1. Differintegrals of $\ln(x)$ for q in the range -2 to $+2$; $d^q\ln(x)/dx^q = x^{-q}[\ln(x) - \gamma - \psi(1-q)]/\Gamma(1-q)$.

Equation (6.7.1) may be regarded as a special instance of the more general result (whose proof we omit)

(6.7.2) $$\frac{d^q[x^p\ln(x)]}{dx^q} = \frac{\Gamma(p+1)x^{p-q}}{\Gamma(p-q+1)}\left[\ln(x) + \psi(p+1) - \psi(p-q+1)\right]$$

for $p > -1$ and $q < 0$.

Using the latter result it is possible to derive the formula

$$\frac{d^q}{dx^q}\frac{d^Q\ln(x)}{dx^Q} = \frac{d^{q+Q}\ln(x)}{dx^{q+Q}},$$

at least for $Q < 1$, and thereby establish the composition rule for the logarithm function for certain ranges of Q and q. It will be recalled that in Section 5.7 the composition rule was derived only for the class of functions comprised by all differintegrable series, a class from which $\ln(x)$ is excluded.

[2] From the limit of $\psi(1-q)/\Gamma(1-q)$ as q approaches the positive integer n, the rule $d^n\ln(x)/dx^n = -\Gamma(n)[-x]^{-n}$ is readily deduced.

6.8 THE HEAVISIDE AND DIRAC FUNCTIONS

The Heaviside function (or unit-step function) occurring at $x = x_0$,

$$H(x - x_0) \equiv \begin{cases} 0, & x < x_0, \\ 1, & x > x_0, \end{cases}$$

is the simplest example of a piecewise-defined function, the general class of which is considered in the next section. The Dirac function (or delta function) $\delta(x - x_0)$ is the derivative of the Heaviside function: It is everywhere zero except at $x = x_0$, where it is infinite.

The differintegration of the Heaviside function for $q < 0$ and with $a < x_0 < x$ is a trivial operation via the Riemann–Liouville definition (3.2.3). Thus

$$(6.8.1) \quad \frac{d^q}{[d(x-a)]^q} H(x - x_0) = \frac{1}{\Gamma(-q)} \int_a^x \frac{H(y - x_0)\, dy}{[x - y]^{q+1}}, \quad q < 0$$

$$= \frac{1}{\Gamma(-q)} \int_a^{x_0} \frac{[0]\, dy}{[x - y]^{q+1}} + \frac{1}{\Gamma(-q)} \int_{x_0}^x \frac{[1]\, dy}{[x - y]^{q+1}}$$

$$= 0 + \frac{d^q[1]}{[d(x - x_0)]^q}, \quad x > x_0$$

$$= \frac{[x - x_0]^{-q}}{\Gamma(1 - q)}, \quad x > x_0,$$

where the results of Sections 4.2 and 4.1 have been employed. By invoking equation (3.2.5) it is easily demonstrated that the same equation applies for all q. Figure 6.8.1 illustrates this formula.

One of the principal uses of the Heaviside function is to delimit the range of definition of a function f. Thus, we have the product

$$f H(x - x_0) = \begin{cases} 0, & x < x_0, \\ f, & x > x_0. \end{cases}$$

We shall need to utilize differintegrals of such a product. By analogy with the derivation (6.8.1) we easily establish that

$$(6.8.2) \quad \frac{d^q}{[d(x-a)]^q} \{f H(x - x_0)\} = H(x - x_0) \frac{d^q f}{[d(x - x_0)]^q},$$

where $a < x_0 < x$.

We characterize the Dirac delta function $\delta(x - x_0)$ by means of its property

$$\int_a^x \delta(y - x_0) f(y)\, dy = f(x_0), \quad a < x_0 < x,$$

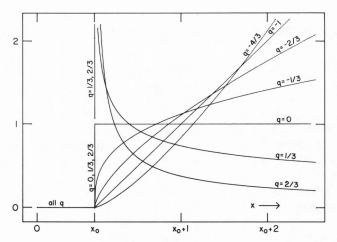

FIG. 6.8.1. Some differintegrals of the Heaviside function $H(x - x_0)$. For $q = 1$ differintegration yields the ungraphable delta function and the $q > 1$ differintegrals are likewise impossible to depict.

for any function f. From this we see immediately on selecting $f = [x - y]^{-q-1}$ that

$$(6.8.3) \qquad \int_a^x \frac{\delta(y - x_0)\, dy}{[x - y]^{q+1}} = [x - x_0]^{-q-1}, \qquad q < 0.$$

If we divide equation (6.8.3) by $\Gamma(-q)$ we recognize the left-hand side as the qth differintegral of the Dirac function; that is,

$$(6.8.4) \qquad \frac{d^q \delta(x - x_0)}{[d(x - a)]^q} = \frac{[x - x_0]^{-q-1}}{\Gamma(-q)}.$$

Extension of formula (6.8.4) to negative q is once again accomplished via equation (3.2.5).

Comparing formulas (6.8.1) and (6.8.4) it will be noticed that

$$\frac{d^{q+1} H(x - x_0)}{[d(x - a)]^{q+1}} = \frac{d^q \delta(x - x_0)}{[d(x - a)]^q}.$$

The well-known relationship

$$\frac{d}{dx} H(x - x_0) = \delta(x - x_0)$$

follows upon setting $q = 0$.

6.9 THE SAWTOOTH FUNCTION

In this section we will first derive a general formula for the differintegral of a piecewise-defined function, and then apply the formula to a simple example, the sawtooth function.

Consider first the function

$$f = \begin{cases} f_1, & a \leq x < x_1, \\ f_2, & x_1 < x. \end{cases}$$

Making use of the Heaviside function, we may write f as

$$f = f_1 H(x_1 - x) + f_2 H(x - x_1) = f_1 - f_1 H(x - x_1) + f_2 H(x - x_1)$$
$$= f_1 + [f_2 - f_1] H(x - x_1)$$

since

$$H(x - x_1) + H(x_1 - x) = 1, \quad x \neq x_1.$$

Application of linearity and equation (6.8.2) is now all that is needed to establish

$$(6.9.1) \quad \frac{d^q f}{[d(x-a)]^q} = \frac{d^q f_1}{[d(x-a)]^q} + H(x - x_1) \frac{d^q [f_2 - f_1]}{[d(x-x_1)]^q}, \quad x \neq x_1.$$

The generalization of this result to the many-sectioned piecewise-defined function

$$f = f_k, \quad x_{k-1} < x < x_k, \quad k = 1, 2, 3, \ldots, N,$$
$$= f_1 + [f_2 - f_1] H(x - x_1) + \cdots + [f_N - f_{N-1}] H(x - x_{N-1}), \quad x \neq x_k,$$

(where $x_0 \equiv a$) is straightforward. The general result

$$(6.9.2) \quad \frac{d^q f}{[d(x-a)]^q} = \frac{d^q f_1}{[d(x-a)]^q} + \sum_{k=1}^{N-1} H(x - x_k) \frac{d^q [f_{k+1} - f_k]}{[d(x-x_k)]^q}$$

allows for an arbitrarily large number of sections.

The sawtooth function, defined by

$$\text{saw}(x) = [-]^k [2k - x - 2], \quad 2k - 3 < x < 2k - 1, \quad k = 1, 2, \ldots, N$$
$$= x + \sum_{k=1}^{N-1} [-]^k [2x - 4k + 2] H(x - 2k + 1)$$

is shown in Fig. 6.9.1. Application of formula (6.9.2), with a lower limit of zero, yields

$$\frac{d^q \operatorname{saw}(x)}{dx^q} = \frac{d^q x}{dx^q} + \sum_{k=1}^{N-1} H(x - 2k + 1) \frac{d^q\{[-]^k[2x - 4k + 2]\}}{[d(x - 2k + 1)]^q}$$

$$= \frac{x^{1-q}}{\Gamma(2-q)} + 2 \sum_{k=1}^{N-1} H(x - 2k + 1)[-]^k \frac{[x - 2k + 1]^{1-q}}{\Gamma(2-q)}.$$

Examples of this differintegral are incorporated into Fig. 6.9.1.

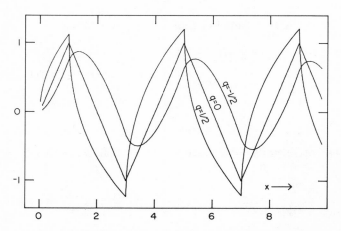

FIG. 6.9.1. The sawtooth function saw(x) and two of its differintegrals.

6.10 PERIODIC FUNCTIONS

Any periodic function is expressible as

(6.10.1) $$\operatorname{per}(x) = \sum_{k=1}^{\infty} \left[C_k \exp\left(\frac{2\pi i k x}{X}\right) + \overline{C}_k \exp\left(\frac{-2\pi i k x}{X}\right) \right],$$

where C_k and $\overline{C}_k$ are conjugate complex constants and X is the period of the function. Differintegration of a periodic function thus devolves into determining

$$\frac{d^q}{dx^q} \exp\left(\frac{\pm 2\pi i k x}{X}\right),$$

an operation to which we now turn our attention.

6.10 PERIODIC FUNCTIONS

Since our proofs of the scale-change theorem (Section 5.4) and of the differintegration of exponential functions (Section 6.2) in no way precluded imaginary constants or variables, we can write

$$(6.10.2) \quad \frac{d^q}{dx^q}\exp\left(\frac{\pm 2\pi ikx}{X}\right) = \left[\frac{\pm 2\pi ik}{X}\right]^q \frac{d^q}{[d(\pm 2\pi ikx/X)]^q}\exp\left(\frac{\pm 2\pi ikx}{X}\right)$$

$$= x^{-q}\exp\left(\frac{\pm 2\pi ikx}{X}\right)\gamma^*\left(-q, \frac{\pm 2\pi ikx}{X}\right).$$

Introduced into (6.10.1), this formula completely describes the differintegration of any periodic function for any value of x. However, it is of interest to find the limiting forms of the differintegral corresponding to small and to large x values.

For small x values a power series of the incomplete gamma function permits the rewriting of equation (6.10.2) as

$$\frac{d^q}{dx^q}\exp\left(\frac{\pm 2\pi ikx}{X}\right) = x^{-q}\sum_{j=0}^{\infty}\frac{[\pm 2\pi ikx/X]^j}{\Gamma(j-q+1)}$$

so that, in the limit of small x, the differintegral of a periodic function is

$$\frac{d^q}{dx^q}\mathrm{per}(x \to 0) = \sum_{k=1}^{\infty}\left[\frac{[C_k + \bar{C}_k]x^{-q}}{\Gamma(1-q)} + \frac{i[C_k - \bar{C}_k]kx^{1-q}}{X\Gamma(2-q)} + \cdots\right].$$

Notice that the coefficients $[C_k + \bar{C}_k]$ and $i[C_k - \bar{C}_k]$ are real and that the leading term in this expansion is simply the differintegral of the initial value per(0) of the periodic function, treated as a constant.

Of more interest is the limiting form for large x. An asymptotic expansion (Abramowitz and Stegun, 1964, p. 263) of the incomplete gamma function leads to

$$\frac{d^q}{dx^q}\exp\left(\frac{\pm 2\pi ikx}{X}\right) \sim \left[\frac{\pm 2\pi ik}{X}\right]^q \exp\left(\frac{\pm 2\pi ikx}{X}\right) - \sum_{j=0}^{\infty}\frac{[\pm 2\pi ikx/X]^{-1-j}x^q}{\Gamma(-q-j)}$$

as equivalent to (6.10.2) in the limit as x tends to infinity. Because $(\pm i)^q$ are the complex numbers $\exp(\pm \pi i q/2)$, this expansion yields

$$\frac{d^q}{dx^q}\mathrm{per}(x \to \infty)$$

$$= \sum_{k=1}^{\infty}\left[\frac{2\pi k}{X}\right]^q\left\{C_k \exp\left(2\pi i\left[\frac{kx}{X} + \frac{q}{4}\right]\right) + \bar{C}_k \exp\left(-2\pi i\left[\frac{kx}{X} + \frac{q}{4}\right]\right)\right\}$$

$$+ \sum_{k=1}^{\infty}\left\{\frac{i[C_k - \bar{C}_k]x^{-1-q}X}{2\pi k\Gamma(-q)} + \frac{[C_k + \bar{C}_k]x^{-2-q}X^2}{4\pi^2 k^2 \Gamma(-q-1)} + \cdots\right\}$$

upon incorporation into the general expression for the differintegral of a periodic function. The terms grouped within the first summation are periodic: They show that the effect of differintegration to order q has been to change the amplitude of each component of the original function by a factor of $[2\pi k/X]^q$ and to change its phase by an angle $\pi q/2$ radians. Within the second summation the terms are aperiodic: Provided q exceeds -1, they represent transients which eventually (i.e., at large x) become insignificant.

The simplest periodic functions are $\sin(x)$ and $\cos(x)$. In the x approaching infinity limit, the differintegrals of these functions are

$$(6.10.3) \qquad \frac{d^q}{dx^q}\sin(x) = \sin\left(x + \frac{\pi q}{2}\right) + \frac{x^{-1-q}}{\Gamma(-q)} - \frac{x^{-3-q}}{\Gamma(-q-2)} + \cdots$$

and

$$(6.10.4) \qquad \frac{d^q}{dx^q}\cos(x) = \cos\left(x + \frac{\pi q}{2}\right) + \frac{x^{-2-q}}{\Gamma(-q-1)} - \frac{x^{-4-q}}{\Gamma(-q-3)} + \cdots.$$

Notice how these equations reduce to well-known formulas in the classical cases of positive or negative integer q.

6.11 CYCLODIFFERENTIAL FUNCTIONS

By this term we mean functions such as $\exp(\pm x)$, $\cosh(x)$, and $\sin(x)$, which are regenerated after sufficient differentiations. In Section 7.6 we shall seek functions which are preserved under repeated semidifferentiation, but for now we restrict consideration to regeneration after a small number of integer order differentiations.

Formula (3.6.5) provides a means of differintegrating cyclodifferential functions which is unusually powerful in that a nonzero lower limit may be employed even with relatively complex functions. We illustrate this by setting $\phi(x) = \sin(x)$ in formula (3.6.5), whereby

$$\frac{d^q \sin(x)}{[d(x-a)]^q} = \frac{1}{\Gamma(-q)} \left\{ \sin(x) \sum_{k=0,4,\ldots} - \cos(x) \sum_{k=1,5,\ldots} - \sin(x) \sum_{k=2,6,\ldots} \right.$$

$$\left. + \cos(x) \sum_{k=3,7,\ldots} \right\} \frac{[x-a]^{k-q}}{k!\,[k-q]}$$

6.11 CYCLODIFFERENTIAL FUNCTIONS

is obtained. This formulation is intended to convey that the same summand, $[x-a]^{k-q}/\{k![k-q]\}$, is summed using four different k-sequences. On regrouping terms,

$$\frac{d^q \sin(x)}{[d(x-a)]^q} = \frac{\sin(x)}{\Gamma(-q)} \sum_{j=0,1,\ldots}^{\infty} \frac{[-]^j[x-a]^{2j-q}}{(2j)![2j-q]}$$
$$- \frac{\cos(x)}{\Gamma(-q)} \sum_{j=0,1,\ldots}^{\infty} \frac{[-]^j[x-a]^{2j-q+1}}{(2j+1)![2j-q+1]}$$

results. At this stage we could proceed to express the summations as hypergeometric functions. However, we shall instead identify them as the indefinite integrals,

(6.11.1) $\quad \dfrac{d^q \sin(x)}{[d(x-a)]^q} = \dfrac{\sin(x)}{\Gamma(-q)} \displaystyle\int_0^{x-a} \dfrac{\cos(u)}{u^{q+1}}\, du - \dfrac{\cos(x)}{\Gamma(-q)} \displaystyle\int_0^{x-a} \dfrac{\sin(u)}{u^{q+1}}\, du,$

provided $q < 0$. To cover the $q < 2$ range, we can withdraw the $j = 0$ terms from the summation, leading to

$$\frac{d^q \sin(x)}{[d(x-a)]^q} = \frac{\sin(x)}{\Gamma(-q)} \left[\frac{[x-a]^{-q}}{-q} + \int_0^{x-a} \frac{\cos(u)-1}{u^{q+1}}\, du \right]$$
$$- \frac{\cos(x)}{\Gamma(-q)} \left[\frac{[x-a]^{1-q}}{1-q} + \int_0^{x-a} \frac{\sin(u)-u}{u^{q+1}}\, du \right].$$

Notice that by defining $u = x - y$, equation (6.11.1) could have been generated directly from the Riemann–Liouville definition

$$\frac{d^q \sin(x)}{[d(x-a)]^q} = \frac{1}{\Gamma(-q)} \int_a^x \frac{\sin(y)\, dy}{[x-y]^{q+1}}$$

of the differintegral of $\sin(x)$. It is unusual to be able to remove the upper differintegration limit x from within the Riemann–Liouville integral in this way. It is possible in the case $f(x) = \sin(x)$ because $\sin(x-u)$ is expressible as a sum of products $F(x)G(u)$ with separated variables. Our discussion here suggests that the ability of a function $f(x-u)$ to be factored in this way to $\sum F(x)G(u)$ is an inherent property of a cyclodifferential function. A search for corresponding algebraic properties in functions (see Section 7.6) that play an analogous role to cyclodifferentials in the semicalculus, however, proved unsuccessful.

The integrals in equation (6.11.1), and in the corresponding equation for the differintegration of the cosine function, are not evaluable in terms of established functions (except for special q values), but they are expressible as generalized $\frac{1}{3}$ hypergeometric functions. The formulas

$$\frac{d^q \sin(x)}{[d(x-a)]^q} = \frac{\sqrt{\pi} \sin(x)}{2\Gamma(-q)[x-a]^q} \left[-\tfrac{1}{4}[x-a]^2 \; \frac{-\tfrac{1}{2}q-1}{-\tfrac{1}{2}q, -\tfrac{1}{2}, 0} \right]$$

$$- \frac{\sqrt{\pi} \cos(x)}{2\Gamma(-q)[x-a]^q} \left[-\tfrac{1}{4}[x-a]^2 \; \frac{-\tfrac{1}{2}q-\tfrac{1}{2}}{-\tfrac{1}{2}q+\tfrac{1}{2}, 0, \tfrac{1}{2}} \right]$$

and

$$\frac{d^q \cos(x)}{[d(x-a)]^q} = \frac{\sqrt{\pi} \cos(x)}{2\Gamma(-q)[x-a]^q} \left[-\tfrac{1}{4}[x-a]^2 \; \frac{-\tfrac{1}{2}q-1}{-\tfrac{1}{2}q, -\tfrac{1}{2}, 0} \right]$$

$$+ \frac{\sqrt{\pi} \sin(x)}{2\Gamma(-q)[x-a]^q} \left[-\tfrac{1}{4}[x-a]^2 \; \frac{-\tfrac{1}{2}q-\tfrac{1}{2}}{-\tfrac{1}{2}q+\tfrac{1}{2}, 0, \tfrac{1}{2}} \right]$$

thereby emerge for $q < 0$.

6.12 THE FUNCTION $x^{q-1} \exp[-1/x]$

In the previous section we noted that the exponential function, in common with only a few other special functions f, possesses the property

$$f(x \pm u) = \sum_i F_i(x) G_i(u),$$

by which an argument that is a linear combination of two variables can be decomposed to a finite sum of products. This property also permits the function $x^{q-1} \exp(-1/x)$ to be differintegrated to order q with a zero lower limit, as we now demonstrate.

Using the Riemann–Liouville definition followed by the substitution $y = x/[xz+1]$, we find

$$\frac{d^q}{dx^q}\left\{\frac{\exp(-1/x)}{x^{1-q}}\right\} = \frac{1}{\Gamma(-q)} \int_0^x \frac{y^{q-1} \exp(-1/y) \, dy}{[x-y]^{q+1}}$$

$$= \frac{\exp(-1/x)}{\Gamma(-q) x^{q+1}} \int_0^\infty \frac{\exp(-z) dz}{z^{q+1}}.$$

From definition (1.3.1), the integral is evaluated simply as $\Gamma(-q)$ so that the final result

(6.12.1) $$\frac{d^q}{dx^q}\left\{\frac{\exp(-1/x)}{x^{1-q}}\right\} = \frac{\exp(-1/x)}{x^{q+1}}$$

emerges. We shall omit the proof of the more general result

$$\frac{d^q}{dx^q}\left\{x^{q-n}\exp\left(\frac{-1}{x}\right)\right\} = \exp\left(\frac{-1}{x}\right)\sum_{j=0}^{n-1}\frac{\Gamma(j-q)}{\Gamma(-q)}\binom{n-1}{j}x^{j-n-q}, n = 1, 2, 3, \ldots$$

of which equation (6.12.1) is the simplest instance.

The corresponding trigonometric differintegrals are

$$\frac{d^q}{dx^q}\left\{x^{q-1}\sin\left(\frac{1}{x}\right)\right\} = x^{-q-1}\sin\left(\frac{1}{x} - \frac{q\pi}{2}\right)$$

and

$$\frac{d^q}{dx^q}\left\{x^{q-1}\cos\left(\frac{1}{x}\right)\right\} = x^{-q-1}\cos\left(\frac{1}{x} - \frac{q\pi}{2}\right).$$

The proof of these results, for $0 < q < 1$, is established by an argument very similar to that of the previous paragraph. It is of interest to note that differintegration of these trigonometric functions has introduced a "phase shift" of $\pi q/2$, as was found in equations (6.10.3) and (6.10.4), but that the shift is in the opposite sense.

CHAPTER 7

SEMIDERIVATIVES AND SEMIINTEGRALS

Within our experience, the most useful applications of derivatives and integrals of noninteger order involve a zero lower limit and moiety order (see Chapter 11). The present chapter, therefore, is devoted to the operators $d^{1/2}/dx^{1/2}$ and $d^{-1/2}/dx^{-1/2}$. We shall here present the effect of these operators on a large variety of functions and also some general properties of these special operators. The presentation is in the form of an extended table, no proofs being given. Generally, the formulas quoted are either straightforward specializations of those given in Chapters 3–6, or can be derived by application of the rules of Chapter 5. Though the chapter is lengthy, it by no means exhausts the possible operands to which semidifferentiation and semiintegration[1] may be applied. We know of no preexisting tabulation of semiderivatives or semiintegrals, though a table of Riemann–Liouville transforms (Erdélyi et al., 1954) can, with caution, be adapted to this purpose. To a limited extent, tables of Laplace transforms (for example, Roberts and Kaufman, 1966) are useful in constructing tables of semiderivatives and semiintegrals, as a result of the transformation properties we shall discuss in Section 8.1.

Though the tables which follow have been designed to have their maximum utility when x has positive real values, most of the entries are valid more generally.

7.1 DEFINITIONS

In this section we specialize to $a = 0$, $q = \pm\frac{1}{2}$ the differintegral definitions summarized in Section 3.6. This specialization leads directly to all table

---

[1] The names "semidifferentiation" and "semiintegration" seem natural to denote the operations performed by the $d^{1/2}/dx^{1/2}$ and $d^{-1/2}/dx^{-1/2}$ operators. Likewise we shall term $d^{1/2}f/dx^{1/2}$ the "semiderivative" of f, and $d^{-1/2}f/dx^{-1/2}$ its "semiintegral." Less frequently we will use "sesquiderivative" of f to indicate $d^{3/2}f/dx^{3/2}$, and similarly for $d^{-3/2}f/dx^{-3/2}$, etc.

7 SEMIDERIVATIVES AND SEMIINTEGRALS

entries except the fourth. The latter is derived making use of the composition rule.

		$\dfrac{d^{\frac{1}{2}}f}{dx^{\frac{1}{2}}}$	$\dfrac{d^{-\frac{1}{2}}f}{dx^{-\frac{1}{2}}}$	
f				
f		$\lim\limits_{N\to\infty}\left\{\sqrt{\dfrac{N}{x}}\sum\limits_{j=0}^{N-1}\dfrac{(2j)!f\left(x-\dfrac{jx}{N}\right)}{[2j-1][2^j j!]^2}\right\}$	$\lim\limits_{N\to\infty}\left\{\sqrt{\dfrac{x}{N}}\sum\limits_{j=0}^{N-1}\dfrac{(2j)!f\left(x-\dfrac{jx}{N}\right)}{[2^j j!]^2}\right\}$	
f		$\dfrac{d}{dx}\left\{\dfrac{d^{-\frac{1}{2}}f}{dx^{-\frac{1}{2}}}\right\}$	$\dfrac{1}{\sqrt{\pi}}\int_0^x \dfrac{f(y)\,dy}{\sqrt{x-y}}$	
f		$\dfrac{f(0)}{\sqrt{\pi x}}+\dfrac{1}{\sqrt{\pi}}\int_0^x \dfrac{f^{(1)}(y)\,dy}{\sqrt{x-y}}$	$\dfrac{1}{\sqrt{\pi}}\int_0^x \dfrac{f(y)\,dy}{\sqrt{x-y}}$	
f		$\dfrac{f(x)}{\sqrt{\pi x}}+\dfrac{1}{2\sqrt{\pi}}\int_0^x \dfrac{[f(x)-f(y)]\,dy}{[x-y]^{\frac{3}{2}}}$	$\dfrac{F(x)-F(0)}{\sqrt{\pi x}}+\dfrac{1}{2\sqrt{\pi}}\int_0^x \dfrac{[F(x)-F(y)]\,dy}{[x-y]^{\frac{3}{2}}}$, $f(y)=\dfrac{dF}{dy}$	
ϕ		$\sum\limits_{k=0}^{\infty}\dfrac{[-x]^k \phi^{(k)}}{\sqrt{\pi x}\,[1-2k]k!}$	$\sum\limits_{k=0}^{\infty}\sqrt{\dfrac{x}{\pi}}\dfrac{[-x]^k \phi^{(k)}}{[k+\frac{1}{2}]k!}$	
ϕ		$\sum\limits_{k=0}^{\infty}\dfrac{x^{k-\frac{1}{2}}\phi^{(k)}(0)}{\Gamma(k+\frac{1}{2})}$	$\sum\limits_{k=0}^{\infty}\dfrac{x^{k+\frac{1}{2}}\phi^{(k)}(0)}{\Gamma(k+\frac{3}{2})}$	

In the preceding table, as in the following section, we use ϕ and ψ to represent real analytic functions, and f an arbitrary differintegrable function of x.

7.2 GENERAL PROPERTIES

In this section we shall specialize the findings of Chapter 5 to the cases $a=0$, $q=\pm\frac{1}{2}$. The same sequence is followed, so that little commentary is needed to supplement the table.

The second and third entries, exemplifying respectively term-by-term differintegration of analytic functions and arbitrary differintegrable series, follow directly from the results of Section 5.2. The eleventh entry, which stems from the composition rule, requires both f and $d^q f/dx^q$ to be differintegrable series.

f	$\dfrac{d^{\frac{1}{2}}f}{dx^{\frac{1}{2}}}$	$\dfrac{d^{-\frac{1}{2}}f}{dx^{-\frac{1}{2}}}$
$f_1 \pm f_2$	$\dfrac{d^{\frac{1}{2}}f_1}{dx^{\frac{1}{2}}} \pm \dfrac{d^{\frac{1}{2}}f_2}{dx^{\frac{1}{2}}}$	$\dfrac{d^{-\frac{1}{2}}f_1}{dx^{-\frac{1}{2}}} \pm \dfrac{d^{-\frac{1}{2}}f_2}{dx^{-\frac{1}{2}}}$
$\sum_{j=0}^{\infty} \phi_j$	$\sum_{j=0}^{\infty} \dfrac{d^{\frac{1}{2}}\phi_j}{dx^{\frac{1}{2}}}$	$\sum_{j=0}^{\infty} \dfrac{d^{-\frac{1}{2}}\phi_j}{dx^{-\frac{1}{2}}}$
$x^p \sum_{j=0}^{\infty} a_j x^j,$ $p > -1$	$\sum_{j=0}^{\infty} a_j \dfrac{\Gamma(p+j+1)}{\Gamma(p+j+\frac{1}{2})} x^{p+j-\frac{1}{2}}$	$\sum_{j=0}^{\infty} a_j \dfrac{\Gamma(p+j+1)}{\Gamma(p+j+\frac{3}{2})} x^{p+j+\frac{1}{2}}$
Cf	$C \dfrac{d^{\frac{1}{2}}f}{dx^{\frac{1}{2}}}$	$C \dfrac{d^{-\frac{1}{2}}f}{dx^{-\frac{1}{2}}}$
$f(\beta x)$	$\sqrt{\beta}\, \dfrac{d^{\frac{1}{2}}}{[d(\beta x)]^{\frac{1}{2}}} f(\beta x)$	$\dfrac{1}{\sqrt{\beta}} \dfrac{d^{-\frac{1}{2}}}{[d(\beta x)]^{-\frac{1}{2}}} f(\beta x)$
$f(-x)$	$i \dfrac{d^{\frac{1}{2}}}{[d(-x)]^{\frac{1}{2}}} f(-x)$	$-i \dfrac{d^{-\frac{1}{2}}}{[d(-x)]^{-\frac{1}{2}}} f(-x)$
$\phi\psi$	$\sum_{j=0}^{\infty} \binom{\frac{1}{2}}{j} \dfrac{d^{\frac{1}{2}-j}\phi}{dx^{\frac{1}{2}-j}} [\psi^{(j)}]$	$\sum_{j=0}^{\infty} \binom{-\frac{1}{2}}{j} \dfrac{d^{-\frac{1}{2}-j}\phi}{dx^{-\frac{1}{2}-j}} [\psi^{(j)}]$
$f p_n$	$\sum_{j=0}^{n} \binom{\frac{1}{2}}{j} \dfrac{d^{\frac{1}{2}-j}f}{dx^{\frac{1}{2}-j}} [p_n^{(j)}]$	$\sum_{j=0}^{n} \binom{-\frac{1}{2}}{j} \dfrac{d^{-\frac{1}{2}-j}f}{dx^{-\frac{1}{2}-j}} [p_n^{(j)}]$
xf	$x \dfrac{d^{\frac{1}{2}}f}{dx^{\frac{1}{2}}} + \dfrac{1}{2} \dfrac{d^{-\frac{1}{2}}f}{dx^{-\frac{1}{2}}}$	$x \dfrac{d^{-\frac{1}{2}}f}{dx^{-\frac{1}{2}}} - \dfrac{1}{2} \dfrac{d^{-\frac{3}{2}}f}{dx^{-\frac{3}{2}}}$
$x^n f$	$\dfrac{\sqrt{\pi}}{2} \sum_{j=0}^{n} \binom{n}{j} \dfrac{x^{n-j}}{\Gamma(\frac{3}{2}-j)} \dfrac{d^{\frac{1}{2}-j}f}{dx^{\frac{1}{2}-j}}$	$\sqrt{\pi} \sum_{j=0}^{n} \binom{n}{j} \dfrac{x^{n-j}}{\Gamma(\frac{1}{2}-j)} \dfrac{d^{-\frac{1}{2}-j}f}{dx^{-\frac{1}{2}-j}}$
$\dfrac{d^q f}{dx^q}$	$\dfrac{d^{q+\frac{1}{2}}f}{dx^{q+\frac{1}{2}}},\ \dfrac{d^{-q}}{dx^{-q}}\dfrac{d^q f}{dx^q} = f$	$\dfrac{d^{q-\frac{1}{2}}f}{dx^{q-\frac{1}{2}}},\ \dfrac{d^{-q}}{dx^{-q}}\dfrac{d^q f}{dx^q} = f$
$\dfrac{df}{dx}$	$\dfrac{d^{\frac{3}{2}}f}{dx^{\frac{3}{2}}} - \dfrac{x^{-\frac{3}{2}}f(0)}{2\sqrt{\pi}},\ f(0) \neq \infty$	$\dfrac{f(0)}{\sqrt{\pi}} + \dfrac{d^{\frac{1}{2}}f}{dx^{\frac{1}{2}}},\ f(0) \neq \infty$
$\dfrac{d^n f}{dx^n},$ $n = 1, 2, \ldots$	$\dfrac{d^{n+\frac{1}{2}}f}{dx^{n+\frac{1}{2}}} + \sum_{k=0}^{n-1} \dfrac{x^{k-n-\frac{1}{2}}f^{(k)}(0)}{\Gamma(k-n+\frac{1}{2})},$ $f^{(k)}(0) \neq \infty$	$\dfrac{d^{n-\frac{1}{2}}f}{dx^{n-\frac{1}{2}}} + \sum_{k=0}^{n-1} \dfrac{x^{k-n+\frac{1}{2}}f^{(k)}(0)}{\Gamma(k-n+\frac{3}{2})},$ $f^{(k)}(0) \neq \infty$
$\phi(x+A)$	$\dfrac{d^{\frac{1}{2}}}{[d(x+A)]^{\frac{1}{2}}} \phi(x+A)$ $- \sum_{k=1}^{\infty} \dfrac{[x+A]^{-\frac{1}{2}-k}}{\Gamma(\frac{1}{2}-k)} \dfrac{d^{-k}}{dA^{-k}} \phi(A)$	$\dfrac{d^{-\frac{1}{2}}}{[d(x+A)]^{-\frac{1}{2}}} \phi(x+A)$ $- \sum_{k=1}^{\infty} \dfrac{[x+A]^{\frac{1}{2}-k}}{\Gamma(\frac{3}{2}-k)} \dfrac{d^{-k}}{dA^{-k}} \phi(A)$
$f,\ x \to 0$	$\dfrac{f(0)}{\sqrt{\pi x}},\ f(0) \neq 0$	$2f(0)\sqrt{\dfrac{x}{\pi}},\ f(0) \neq 0$

7 SEMIDERIVATIVES AND SEMIINTEGRALS

In the table we have used p_n to represent a polynomial of degree n in x. The binomial coefficients $\binom{\frac{1}{2}}{j}$ and $\binom{-\frac{1}{2}}{j}$ occur several times in the foregoing. Values for many of these coefficients are listed in Table 7.2.1. In certain applications it is the values of finite or infinite sums of $\binom{\frac{1}{2}}{j}$ or $\binom{-\frac{1}{2}}{j}$ which are needed; such values are also to be found in the tabulation.

Table 7.2.1. *Values of* $\binom{\frac{1}{2}}{j}$, $\binom{-\frac{1}{2}}{j}$ *and their cumulative sums.*

J	$\binom{\frac{1}{2}}{j}$	$\sum_{j=0}^{J}\binom{\frac{1}{2}}{j}$	$\binom{-\frac{1}{2}}{j}$	$\sum_{j=0}^{J}\binom{-\frac{1}{2}}{j}$
0	1	1	1	1
1	$\dfrac{1}{2}$	$\dfrac{3}{2}$	$\dfrac{-1}{2}$	$\dfrac{1}{2}$
2	$\dfrac{-1}{8}$	$\dfrac{11}{8}$	$\dfrac{3}{8}$	$\dfrac{7}{8}$
3	$\dfrac{1}{16}$	$\dfrac{23}{16}$	$\dfrac{-5}{16}$	$\dfrac{9}{16}$
4	$\dfrac{-5}{128}$	$\dfrac{179}{128}$	$\dfrac{35}{128}$	$\dfrac{107}{128}$
5	$\dfrac{7}{256}$	$\dfrac{365}{256}$	$\dfrac{-63}{256}$	$\dfrac{151}{256}$
6	$\dfrac{-21}{1024}$	$\dfrac{1439}{1024}$	$\dfrac{231}{1024}$	$\dfrac{835}{1024}$
$\to \infty$	$\dfrac{-[-1]^J}{2\sqrt{\pi}J^{\frac{3}{2}}}$	$\to \sqrt{2}$	$\dfrac{[-1]^J}{\sqrt{\pi J}}$	$\to \dfrac{1}{\sqrt{2}}$

7.3 CONSTANTS AND POWERS

These provide the simplest subjects for semidifferentiation and semiintegration.

7.3 CONSTANTS AND POWERS

f	$\dfrac{d^{\frac{1}{2}}f}{dx^{\frac{1}{2}}}$	$\dfrac{d^{-\frac{1}{2}}f}{dx^{-\frac{1}{2}}}$
0	0	0
C, any constant	$\dfrac{C}{\sqrt{\pi x}}$	$2C\sqrt{\dfrac{x}{\pi}}$
$x^{-\alpha}$, $\alpha = 0.79195\ldots$	$-x^{-\alpha-\frac{1}{2}}$	$\dfrac{x^{\frac{1}{2}-\alpha}}{\alpha-\frac{1}{2}}$
$\dfrac{1}{\sqrt{x}}$	0	$\sqrt{\pi}$
$x^0 \equiv 1$	$\dfrac{1}{\sqrt{\pi x}}$	$2\sqrt{\dfrac{x}{\pi}}$
x^{β}, $\beta = 0.22119\ldots$	$[\beta+\frac{1}{2}]x^{\beta-\frac{1}{2}}$	$x^{\beta+\frac{1}{2}}$
$\sqrt{x}$	$\frac{1}{2}\sqrt{\pi}$	$\frac{1}{2}\sqrt{\pi}\, x$
$x^{\beta+\frac{1}{2}}$	x^{β}	$\dfrac{x^{\beta+1}}{\beta+1}$
x	$2\sqrt{\dfrac{x}{\pi}}$	$\dfrac{4x^{\frac{3}{2}}}{3\sqrt{\pi}}$
$x^{\frac{3}{2}}$	$\frac{3}{4}\sqrt{\pi}\, x$	$\frac{3}{8}\sqrt{\pi}\, x^2$
x^2	$\dfrac{8x^{\frac{3}{2}}}{3\sqrt{\pi}}$	$\dfrac{16x^{\frac{5}{2}}}{15\sqrt{\pi}}$
x^n, $n = 0, 1, 2, \ldots$	$\dfrac{[n!]^2[4x]^n}{(2n)!\sqrt{\pi x}}$	$\dfrac{[n!]^2[4x]^{n+\frac{1}{2}}}{(2n+1)!\sqrt{\pi}}$
$x^{n+\frac{1}{2}}$	$\dfrac{(2n+1)!\sqrt{\pi}}{2[n!]^2}\left[\dfrac{x}{4}\right]^n$	$\dfrac{(2n+2)!\sqrt{\pi}}{[(n+1)!]^2}\left[\dfrac{x}{4}\right]^{n+1}$
x^p, $p > -1$	$\dfrac{\Gamma(p+1)}{\Gamma(p+\frac{1}{2})}x^{p-\frac{1}{2}}$	$\dfrac{\Gamma(p+1)}{\Gamma(p+\frac{3}{2})}x^{p+\frac{1}{2}}$

The exponents α and β appear in the solution of the equations

$$\sqrt{x}\,\frac{d^{\frac{1}{2}}f}{dx^{\frac{1}{2}}} \pm f = 0,$$

which are two of the simplest semidifferential equations (see Section 8.5 for a definition of a semidifferential equation and for more complex examples).

7.4 BINOMIALS

By this title we mean powers of $1 \pm x$; we also include instances of powers of $1 \pm x$ multiplied by powers of x. The validity of the semidifferentiation and semiintegration formulas in this section are generally restricted to $0 < x < 1$, though $x = 1$ is often permitted.

Notice in the table below that the same entry, $\sqrt{\pi}/\{2[1-x]^{3/2}\}$, occurs in the semiderivative column opposite two distinct operands, $1/\{\sqrt{x}[1-x]\}$ and $\sqrt{x}/[1-x]$. This result is correct, though it may appear questionable. The final paragraphs of Section 5.7, which discuss the occasional failure of the $d^{-q}\{d^q f/dx^q\}/dx^{-q} \equiv f$ "rule," provide the key to the apparent anomaly. The paradox is paralleled in the classical calculus by the observation that two distinct functions, say x^2 and $x^2 - 1$, may have the same derivative.

f	$\dfrac{d^{\frac{1}{2}}f}{dx^{\frac{1}{2}}}$	$\dfrac{d^{-\frac{1}{2}}f}{dx^{-\frac{1}{2}}}$
$\sqrt{1+x}$	$\dfrac{1}{\sqrt{\pi x}} + \dfrac{\arctan(\sqrt{x})}{\sqrt{\pi}}$	$\sqrt{\dfrac{x}{\pi}} + \dfrac{[1+x]\arctan(\sqrt{x})}{\sqrt{\pi}}$
$\sqrt{1-x}$	$\dfrac{1}{\sqrt{\pi x}} - \dfrac{\operatorname{arctanh}(\sqrt{x})}{\sqrt{\pi}}$	$\sqrt{\dfrac{x}{\pi}} + \dfrac{[1-x]\operatorname{arctanh}(\sqrt{x})}{\sqrt{\pi}}$
$\dfrac{1}{\sqrt{1+x}}$	$\dfrac{1}{\sqrt{\pi x}[1+x]}$	$\dfrac{2}{\sqrt{\pi}} \arctan(\sqrt{x})$
$\dfrac{1}{\sqrt{1-x}}$	$\dfrac{1}{\sqrt{\pi x}[1-x]}$	$\dfrac{2}{\sqrt{\pi}} \operatorname{arctanh}(\sqrt{x})$
$\dfrac{1}{1+x}$	$\dfrac{\sqrt{1+x} - \sqrt{x}\operatorname{arcsinh}(\sqrt{x})}{\sqrt{\pi x}[1+x]^{3/2}}$	$\dfrac{2\operatorname{arcsinh}(\sqrt{x})}{\sqrt{\pi}[1+x]}$
$\dfrac{1}{1-x}$	$\dfrac{\sqrt{1-x} + \sqrt{x}\arcsin(\sqrt{x})}{\sqrt{\pi x}[1-x]^{3/2}}$	$\dfrac{2\arcsin(\sqrt{x})}{\sqrt{\pi}[1-x]}$
$\dfrac{1}{[1 \pm x]^{3/2}}$	$\dfrac{1 \mp x}{\sqrt{\pi x}[1 \pm x]^2}$	$\dfrac{2}{1 \pm x}\sqrt{\dfrac{x}{\pi}}$

7.4 BINOMIALS

f	$\dfrac{d^{\frac{1}{2}}f}{dx^{\frac{1}{2}}}$	$\dfrac{d^{-\frac{1}{2}}f}{dx^{-\frac{1}{2}}}$
$\dfrac{1}{[1-x]^p}$	$\dfrac{-\mathbf{B}_x(-\frac{1}{2}, p+\frac{1}{2})}{2\sqrt{\pi}[1-x]^{p+\frac{1}{2}}}$	$\dfrac{\mathbf{B}_x(\frac{1}{2}, p-\frac{1}{2})}{\sqrt{\pi}[1-x]^{p-\frac{1}{2}}}$
$\dfrac{1}{\sqrt{x[1-x]}}$	$\dfrac{E(x) - [1-x]K(x)}{x[1-x]\sqrt{\pi}}$	$\dfrac{2}{\sqrt{\pi}}K(x)$
$\dfrac{1}{\sqrt{x[1+x]}}$	$\dfrac{E\!\left(\dfrac{x}{1+x}\right) - K\!\left(\dfrac{x}{1+x}\right)}{x\sqrt{\pi[1+x]}}$	$\dfrac{2K\!\left(\dfrac{x}{1+x}\right)}{\sqrt{\pi[1+x]}}$
$\sqrt{\dfrac{x}{1-x}}$	$\dfrac{E(x)}{[1-x]\sqrt{\pi}}$	$\dfrac{2}{\sqrt{\pi}}[K(x) - E(x)]$
$\{x[1+x]\}^p,$ $\frac{1}{2} - 2\lambda = p > -1$	$\dfrac{\Gamma(p+1)}{\{x[1+x]\}^{\lambda}} P_p^{2\lambda}(2x-1)$	$\dfrac{\Gamma(p+1)}{\{x[1+x]\}^{\lambda-\frac{1}{2}}} P_p^{2\lambda-1}(2x-1)$
$\dfrac{1}{\sqrt{x[1 \pm x]}}$	$\dfrac{\mp\sqrt{\pi}}{2[1 \pm x]^{\frac{3}{2}}}$	$\sqrt{\dfrac{\pi}{1 \pm x}}$
$\dfrac{\sqrt{x}}{\pm x}$	$\dfrac{\sqrt{\pi}}{2[1 \pm x]^{\frac{1}{2}}}$	$\mp\sqrt{\dfrac{\pi}{1 \pm x}} \pm \sqrt{\pi}$
$\dfrac{\sqrt{x}}{[1-x]^2}$	$\dfrac{\sqrt{\pi}[2+x]}{4[1-x]^{\frac{5}{2}}}$	$\dfrac{\sqrt{\pi}x}{2[1-x]^{\frac{3}{2}}}$
$\dfrac{x^p}{[1-x]^{p+\frac{3}{2}}}$	$\dfrac{[p+\frac{1}{2}+\frac{1}{2}x]\Gamma(p+1)x^{p-\frac{1}{2}}}{\Gamma(p+\frac{3}{2})[1-x]^{2+p}}$	$\dfrac{\Gamma(p+1)x^{p+\frac{1}{2}}}{\Gamma(p+\frac{3}{2})[1-x]^{p+1}}$
$\sqrt{\dfrac{1-x}{x}}$	$\dfrac{E(x) - K(x)}{x\sqrt{\pi}}$	$\dfrac{2}{\sqrt{\pi}}E(x)$
$\sqrt{\dfrac{1+x}{x}}$	$\dfrac{[1+x]E\!\left(\dfrac{x}{1+x}\right) - K\!\left(\dfrac{x}{1+x}\right)}{x\sqrt{\pi[1+x]}}$	$2\sqrt{\dfrac{1+x}{\pi}}E\!\left(\dfrac{x}{1+x}\right)$
$\dfrac{x^p}{[1 \pm x]^r},$ $p > -1$	$\dfrac{\Gamma(p+1)}{\Gamma(p+\frac{1}{2})} x^{p-\frac{1}{2}} F(r, p+1; p+\frac{1}{2}; \mp x)$ $\equiv \dfrac{x^{p-\frac{1}{2}}}{\Gamma(-r)}\left[\mp x \dfrac{r-1, p}{0, p-\frac{1}{2}}\right]$	$\dfrac{\Gamma(p+1)}{\Gamma(p+\frac{3}{2})} x^{p+\frac{1}{2}} F(r, p+1; p+\frac{3}{2}; \mp x)$ $\equiv \dfrac{x^{p+\frac{1}{2}}}{\Gamma(-r)}\left[\mp x \dfrac{r-1, p}{0, p+\frac{1}{2}}\right]$

In the preceding table $K(x)$ and $E(x)$ denote the (complete) elliptic integrals of the first

$$K(x) = \int_0^{\pi/2} \frac{d\theta}{\sqrt{1 - x \sin^2(\theta)}}$$

and second

$$E(x) = \int_0^{\pi/2} \sqrt{1 - x \sin^2(\theta)} \, d\theta$$

kinds. The Legendre function of the first kind of degree v, order μ, and argument $y[-1 < y < 1]$ is symbolized here and elsewhere by $P_v^\mu(y)$: Its definition and properties are discussed by Abramowitz and Stegun (1964, Chap. 8). The symbols $B_x(\,,\,)$ and $F(\,,\,;\,;x)$ for the incomplete beta and Gauss hypergeometric functions were introduced in Sections 1.3 and 2.10.

7.5 EXPONENTIAL AND RELATED FUNCTIONS

The exponential function $\exp(x)$ is unique in the classical calculus in being preserved on repeated differentiation with respect to x. On repeated integration with zero lower limit, $\exp(x)$ is again preserved but a constant or power of x is subtracted at each integration. This is a suitable place to enquire if a function of x exists that is preserved on semidifferentiation with zero lower limit, and thus plays a similar role to the exponential in classical calculus. Inspection of the table below shows that

$$\frac{1}{\sqrt{\pi x}} + \exp(x) \operatorname{erfc}(-\sqrt{x})$$

is such a function.[2] Moreover, on repeated semiintegration, this function is preserved but a constant or a power of x is subtracted at each step; thus, the parallel to $\exp(x)$ is complete. Figure 7.5.1 shows this important function and some of its successive semiintegrals.

---

[2] In other words, $[1/\sqrt{\pi x}] + \exp(x)\operatorname{erfc}(-\sqrt{x})$ is an eigenfunction of the $d^{1/2}/dx^{1/2}$ operator. More generally, an eigenfunction of the d^q/dx^q operator ($q > 0$) is the differintegrable series $x^{q-1} \sum C^j x^{jq}/\Gamma([j+1]q)$, where the summation is from $j = 0$ to $j = \infty$ and C is the eigenvalue.

7.5 EXPONENTIAL AND RELATED FUNCTIONS

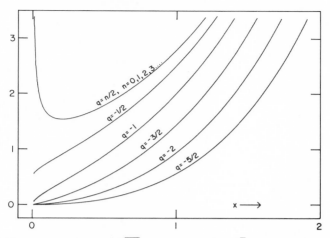

FIG. 7.5.1. The function $[1/\sqrt{\pi x}] + \exp(x)\,\text{erfc}(-\sqrt{x})$, an eigenfunction of the semidifferentiation operator, and some of its successive semiintegrals. This function is preserved by the operator d^q/dx^q for $q = 0, \frac{1}{2}, 1, \frac{3}{2}, 2, \frac{5}{2}$, etc., but repeated semiintegration yields the function illustrated.

f	$\dfrac{d^{\frac{1}{2}}f}{dx^{\frac{1}{2}}}$	$\dfrac{d^{-\frac{1}{2}}f}{dx^{-\frac{1}{2}}}$
$\exp(x)$	$\dfrac{1}{\sqrt{\pi x}} + \exp(x)\,\text{erf}(\sqrt{x})$	$\exp(x)\,\text{erf}(\sqrt{x})$
$\exp(-x)$	$\dfrac{1}{\sqrt{\pi x}} - \dfrac{2}{\sqrt{\pi}}\,\text{daw}(\sqrt{x})$	$\dfrac{2}{\sqrt{\pi}}\,\text{daw}(\sqrt{x})$
$\exp(x)\,\text{erf}(\sqrt{x})$	$\exp(x)$	$\exp(x) - 1$
$\text{daw}(\sqrt{x})$	$\frac{1}{2}\sqrt{\pi}\,\exp(-x)$	$\frac{1}{2}\sqrt{\pi}[1 - \exp(-x)]$
$\exp(x)\,\text{erfc}(\sqrt{x})$	$\dfrac{1}{\sqrt{\pi x}} - \exp(x)\,\text{erfc}(\sqrt{x})$	$1 - \exp(x)\,\text{erfc}(\sqrt{x})$
$\exp(x)\,\text{erfc}(-\sqrt{x})$	$\dfrac{1}{\sqrt{\pi x}} + \exp(x)\,\text{erfc}(-\sqrt{x})$	$\exp(x)\,\text{erfc}(-\sqrt{x}) - 1$
$\text{erf}(\sqrt{x})$	$\exp(-\frac{1}{2}x)I_0(\frac{1}{2}x)$	$x\exp(-\frac{1}{2}x)[I_1(\frac{1}{2}x) + I_0(\frac{1}{2}x)]$
$\exp(\pm x)/\sqrt{x}$	$\frac{1}{2}\sqrt{\pi}\,\exp(\pm\frac{1}{2}x)[I_1(\frac{1}{2}x) \pm I_0(\frac{1}{2}x)]$	$\sqrt{\pi}\,\exp(\pm\frac{1}{2}x)I_0(\frac{1}{2}x)$

table continues

f	$\dfrac{d^{\frac{1}{2}}f}{dx^{\frac{1}{2}}}$	$\dfrac{d^{-\frac{1}{2}}f}{dx^{-\frac{1}{2}}}$
$\dfrac{1}{\sqrt{x}}\exp\left(\dfrac{-1}{x}\right)$	$\dfrac{1}{x^{\frac{3}{2}}}\exp\left(\dfrac{-1}{x}\right)$	$\sqrt{\pi}\,\text{erfc}\left(\dfrac{1}{\sqrt{x}}\right)$
$\text{erfc}\left(\dfrac{1}{\sqrt{x}}\right)$	$\dfrac{1}{\sqrt{\pi x}}\exp\left(\dfrac{-1}{x}\right)$	$2\sqrt{x}\,\text{ierfc}\left(\dfrac{1}{\sqrt{x}}\right)$
$\sqrt{x}\,\exp(x)\gamma^*(c,x)$	$\dfrac{\sqrt{\pi}\,M(\frac{3}{2},c+1,x)}{2\Gamma(c+1)}$	$\dfrac{\sqrt{\pi}}{\Gamma(c)}\,[M(\frac{1}{2},c,x)-1]$

7.6 TRIGONOMETRIC AND HYPERBOLIC FUNCTIONS

As was demonstrated in Section 6.4, the circular and hyperbolic sines of $\sqrt{x}$ give Bessel functions on differintegration. The corresponding cosines similarly yield Struve functions. With argument x, the trigonometric and hyperbolic functions are less easily differintegrated, but a few such results are tabulated below. Some inverse functions are also included.

f	$\dfrac{d^{\frac{1}{2}}f}{dx^{\frac{1}{2}}}$	$\dfrac{d^{-\frac{1}{2}}f}{dx^{-\frac{1}{2}}}$
$\sin(\sqrt{x})$	$\tfrac{1}{2}\sqrt{\pi}\,J_0(\sqrt{x})$	$\sqrt{\pi x}\,J_1(\sqrt{x})$
$\cos(\sqrt{x})$	$\dfrac{1}{\sqrt{\pi x}}-\dfrac{\sqrt{\pi}\,H_0(\sqrt{x})}{2}$	$\sqrt{\pi x}\,H_{-1}(\sqrt{x})$
$\sinh(\sqrt{x})$	$\tfrac{1}{2}\sqrt{\pi}\,I_0(\sqrt{x})$	$\sqrt{\pi x}\,I_1(\sqrt{x})$
$\cosh(\sqrt{x})$	$\dfrac{1}{\sqrt{\pi x}}+\dfrac{\sqrt{\pi}\,L_0(\sqrt{x})}{2}$	$\sqrt{\pi x}\,L_{-1}(\sqrt{x})$
$\dfrac{\sin(\sqrt{x})}{\sqrt{x}}$	$\sqrt{\dfrac{\pi}{x}}\,\dfrac{H_{-1}(\sqrt{x})}{2}$	$\sqrt{\pi}\,H_0(\sqrt{x})$
$\dfrac{\cos(\sqrt{x})}{\sqrt{x}}$	$-\sqrt{\dfrac{\pi}{x}}\,\dfrac{J_1(\sqrt{x})}{2}$	$\sqrt{\pi}\,J_0(\sqrt{x})$
$\dfrac{\sinh(\sqrt{x})}{\sqrt{x}}$	$\sqrt{\dfrac{\pi}{x}}\,\dfrac{L_{-1}(\sqrt{x})}{2}$	$\sqrt{\pi}\,L_0(\sqrt{x})$

7.6 TRIGONOMETRIC AND HYPERBOLIC FUNCTIONS

f	$\dfrac{d^{\frac{1}{2}}f}{dx^{\frac{1}{2}}}$	$\dfrac{d^{-\frac{1}{2}}f}{dx^{-\frac{1}{2}}}$
$\dfrac{\cosh(\sqrt{x})}{\sqrt{x}}$	$\sqrt{\dfrac{\pi}{x}}\,\dfrac{I_1(\sqrt{x})}{2}$	$\sqrt{\pi}\,I_0(\sqrt{x})$
$\dfrac{1-\cos(\sqrt{x})}{x}$	$\dfrac{\sqrt{\pi}}{2}H_0(\sqrt{x})$	$\sqrt{\pi x}\,H_1(\sqrt{x})$
$\dfrac{1-\cosh(\sqrt{x})}{x}$	$\dfrac{\sqrt{\pi}}{2}L_0(\sqrt{x})$	$\sqrt{\pi x}\,L_1(\sqrt{x})$
$\sin(x)$	$\sin(x+\tfrac{1}{4}\pi)$ $-\sqrt{2}\,\mathrm{gres}\left(\sqrt{\dfrac{2x}{\pi}}\right)$	$\sin(x-\tfrac{1}{4}\pi)$ $+\sqrt{2}\,\mathrm{fres}\left(\sqrt{\dfrac{2x}{\pi}}\right)$
$\cos(x)$	$\dfrac{1}{\sqrt{\pi x}}+\cos(x+\tfrac{1}{4}\pi)$ $-\sqrt{2}\,\mathrm{fres}\left(\sqrt{\dfrac{2x}{\pi}}\right)$	$\cos(x-\tfrac{1}{4}\pi)$ $-\sqrt{2}\,\mathrm{gres}\left(\sqrt{\dfrac{2x}{\pi}}\right)$
$\sinh(x)$	$\dfrac{\mathrm{daw}(\sqrt{x})}{\sqrt{\pi}}-\dfrac{\exp(x)\,\mathrm{erf}(\sqrt{x})}{2}$	$\dfrac{\exp(x)\,\mathrm{erf}(\sqrt{x})}{2}-\dfrac{\mathrm{daw}(\sqrt{x})}{\sqrt{\pi}}$
$\cosh(x)$	$\dfrac{1}{\sqrt{\pi x}}+\dfrac{\exp(x)\,\mathrm{erf}(\sqrt{x})}{2}-\dfrac{\mathrm{daw}(\sqrt{x})}{\sqrt{\pi}}$	$\dfrac{\exp(x)\,\mathrm{erf}(\sqrt{x})}{2}+\dfrac{\mathrm{daw}(\sqrt{x})}{\sqrt{\pi}}$
$\dfrac{\arcsin(\sqrt{x})}{\sqrt{1-x}}$	$\dfrac{\sqrt{\pi}}{2[1-x]}$	$-\dfrac{\sqrt{\pi}}{2}\ln(1-x)$
$\arctan(\sqrt{x})$	$\dfrac{1}{2}\sqrt{\dfrac{\pi}{1+x}}$	$\sqrt{\pi}\,[\sqrt{1+x}-1]$
$\dfrac{\mathrm{arcsinh}(\sqrt{x})}{\sqrt{1+x}}$	$\dfrac{\sqrt{\pi}}{2[1+x]}$	$\dfrac{\sqrt{\pi}}{2}\ln(1+x)$
$\mathrm{arctanh}(\sqrt{x})$	$\dfrac{1}{2}\sqrt{\dfrac{\pi}{1-x}}$	$\sqrt{\pi}\,[1-\sqrt{1-x}]$

In the preceding table, $J_\nu(\ )$ and $I_\nu(\ )$ denote respectively the Bessel and modified Bessel functions of order ν; $H_\nu(\ )$ and $L_\nu(\ )$ similarly denote the Struve and modified Struve functions of order ν. The functions we denote fres( ) and gres( ) are those auxiliary Fresnel integrals that Abramowitz and Stegun (1964, p. 300) symbolize as $f(\ )$ and $g(\ )$. These functions occur in the

126 7 SEMIDERIVATIVES AND SEMIINTEGRALS

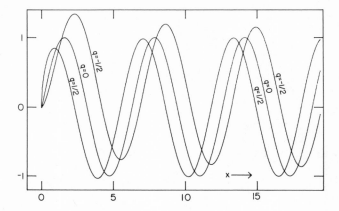

FIG. 7.6.1. The semiderivative ($q = +\frac{1}{2}$) and semiintegral ($q = -\frac{1}{2}$) of sin(x).

semiderivative and semiintegral of the sine (see Fig. 7.6.1) and cosine (see Fig. 7.6.2) functions.

Note that in this section (as, indeed, throughout the book) we have chosen real arguments for all functions. Because of the relationships

$$\sinh(x) = -i \sin(ix),$$
$$\cosh(x) = \cos(ix),$$
$$\sin(x) = \tfrac{1}{2} \exp(-ix) - \tfrac{1}{2} \exp(ix),$$

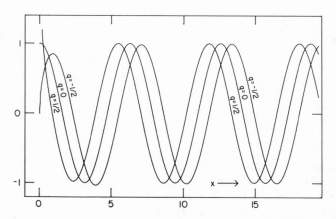

FIG. 7.6.2. The semiderivative ($q = +\frac{1}{2}$) and semiintegral ($q = -\frac{1}{2}$) of cos(x).

etc., it is easy to replace the tabular entries by functions with imaginary arguments, leading to such formulas as

$$\frac{d^{\frac{1}{2}}}{dx^{\frac{1}{2}}} \frac{\sinh(\sqrt{x})}{\sqrt{x}} = -i\sqrt{\frac{\pi}{x}} \frac{H_{-1}(\sqrt{ix})}{2}.$$

7.7 BESSEL AND STRUVE FUNCTIONS

Here we restrict consideration almost solely to the argument $\sqrt{x}$.

f	$\dfrac{d^{\frac{1}{2}}f}{dx^{\frac{1}{2}}}$	$\dfrac{d^{-\frac{1}{2}}f}{dx^{-\frac{1}{2}}}$
$J_0(\sqrt{x})$	$\dfrac{\cos(\sqrt{x})}{\sqrt{\pi x}}$	$\dfrac{2\sin(\sqrt{x})}{\sqrt{\pi}}$
$\dfrac{J_1(\sqrt{x})}{\sqrt{x}}$	$\dfrac{\cos(\sqrt{x}) + \sqrt{x}\sin(\sqrt{x}) - 1}{\sqrt{\pi}x^{\frac{3}{2}}}$	$\dfrac{2[1 - \cos(\sqrt{x})]}{\sqrt{\pi x}}$
$\dfrac{J_\nu(\sqrt{x})}{x^{\nu/2}}$	$\dfrac{1}{2^\nu \Gamma(\nu+1)\sqrt{\pi x}} - \dfrac{H_{\nu+\frac{1}{2}}(\sqrt{x})}{\sqrt{2}x^{[\nu/2]+\frac{1}{4}}}$	$\dfrac{\sqrt{2}H_{\nu-\frac{1}{2}}(\sqrt{x})}{x^{[\nu/2]-\frac{1}{4}}}$
$\sqrt{x}J_1(\sqrt{x})$	$\dfrac{\sin(\sqrt{x})}{\sqrt{\pi}}$	$\dfrac{2\sin(\sqrt{x}) - 2\sqrt{x}\cos(\sqrt{x})}{\sqrt{\pi}}$
$x^{\nu/2}J(\sqrt{x})$	$\dfrac{x^{[\nu/2]-\frac{1}{4}}J_{\nu-\frac{1}{2}}(\sqrt{x})}{\sqrt{2}}$	$\sqrt{2}x^{[\nu/2]+\frac{1}{4}}J_{\nu+\frac{1}{2}}(\sqrt{x})$
$I_0(\sqrt{x})$	$\dfrac{\cosh(\sqrt{x})}{\sqrt{\pi x}}$	$\dfrac{2\sinh(\sqrt{x})}{\sqrt{\pi}}$
$\dfrac{I_1(\sqrt{x})}{\sqrt{x}}$	$\dfrac{\cosh(\sqrt{x}) - \sqrt{x}\sinh(\sqrt{x}) - 1}{\sqrt{\pi}x^{\frac{3}{2}}}$	$\dfrac{2[1 - \cosh(\sqrt{x})]}{\sqrt{\pi x}}$
$\dfrac{I_\nu(\sqrt{x})}{x^{\nu/2}}$	$\dfrac{1}{2^\nu \Gamma(\nu+1)\sqrt{\pi x}} + \dfrac{L_{\nu+\frac{1}{2}}(\sqrt{x})}{\sqrt{2}x^{[\nu/2]+\frac{1}{4}}}$	$\dfrac{\sqrt{2}L_{\nu-\frac{1}{2}}(\sqrt{x})}{x^{[\nu/2]-\frac{1}{4}}}$
$\sqrt{x}I_1(\sqrt{x})$	$\dfrac{\sinh(\sqrt{x})}{\sqrt{\pi}}$	$\dfrac{2\sqrt{x}\cosh(\sqrt{x}) - 2\sinh(\sqrt{x})}{\sqrt{\pi}}$

table continues

7 SEMIDERIVATIVES AND SEMIINTEGRALS

f	$\dfrac{d^{\frac{1}{2}}f}{dx^{\frac{1}{2}}}$	$\dfrac{d^{-\frac{1}{2}}f}{dx^{-\frac{1}{2}}}$
$x^{\nu/2} I_\nu(\sqrt{x})$	$\dfrac{x^{[\nu/2]-\frac{1}{4}} I_{\nu-\frac{1}{2}}(\sqrt{x})}{\sqrt{2}}$	$\sqrt{2}\, x^{[\nu/2]+\frac{1}{4}} I_{\nu+\frac{1}{2}}(\sqrt{x})$
$\exp(-x) I_0(x)$	$\dfrac{\exp(-2x)}{\sqrt{\pi x}}$	$\dfrac{\operatorname{erf}(\sqrt{2x})}{\sqrt{2}}$
$H_0(\sqrt{x})$	$\dfrac{\sin(\sqrt{x})}{\sqrt{\pi x}}$	$\dfrac{2[1-\cos(\sqrt{x})]}{\sqrt{\pi}}$
$\dfrac{H_0(\sqrt{x})}{x}$	$\dfrac{\sin(\sqrt{x}) - \operatorname{Si}(\sqrt{x})}{\sqrt{\pi}\, x^{\frac{3}{2}}}$	$\dfrac{2\operatorname{Si}(\sqrt{x})}{\sqrt{\pi x}}$
$\sqrt{x}\, H_1(\sqrt{x})$	$\dfrac{1-\cos(\sqrt{x})}{\sqrt{\pi}}$	$\dfrac{x + 2 - 2\sqrt{x}\sin(\sqrt{x}) - 2\cos(\sqrt{x})}{\sqrt{\pi}}$
$x^{\nu/2} H_\nu(\sqrt{x})$	$\dfrac{x^{[\nu/2]-\frac{1}{4}} H_{\nu-\frac{1}{2}}(\sqrt{x})}{\sqrt{2}}$	$\sqrt{2}\, x^{[\nu/2]+\frac{1}{4}} H_{\nu+\frac{1}{2}}(\sqrt{x})$
$L_0(\sqrt{x})$	$\dfrac{\sinh(\sqrt{x})}{\sqrt{\pi x}}$	$\dfrac{2[\cosh(\sqrt{x}) - 1]}{\sqrt{\pi}}$
$\dfrac{L_0(\sqrt{x})}{x}$	$\dfrac{\sinh(\sqrt{x}) - \operatorname{Shi}(\sqrt{x})}{\sqrt{\pi}\, x^{\frac{3}{2}}}$	$\dfrac{2\operatorname{Shi}(\sqrt{x})}{\sqrt{\pi x}}$
$\sqrt{x}\, L_1(\sqrt{x})$	$\dfrac{\cosh(\sqrt{x}) - 1}{\sqrt{\pi}}$	$\dfrac{2 - x + 2\sqrt{x}\sinh(\sqrt{x}) - 2\cosh(\sqrt{x})}{\sqrt{\pi}}$
$x^{\nu/2} L_\nu(\sqrt{x})$	$\dfrac{x^{[\nu/2]-\frac{1}{4}} L_{\nu-\frac{1}{2}}(\sqrt{x})}{\sqrt{2}}$	$\sqrt{2}\, x^{[\nu/2]+\frac{1}{4}} L_{\nu+\frac{1}{2}}(\sqrt{x})$

The functions $\operatorname{Si}(\ )$ and $\operatorname{Shi}(\ )$ appearing in this table are the sine integral

$$\operatorname{Si}(x) \equiv \int_0^x \frac{\sin(y)\, dy}{y}$$

and its hyperbolic counterpart [see Abramowitz and Stegun (1964, p. 231)]

$$\operatorname{Shi}(x) \equiv \int_0^x \frac{\sinh(y)\, dy}{y}.$$

7.8 GENERALIZED HYPERGEOMETRIC FUNCTIONS

The results reported here are specializations of the general rules deduced in Section 6.5.

The full range of possible numeratorial and denominatorial parameters has been spelled out in the first six tabular entries; thereafter only ellipses are used to represent their possible presence.

f	$\dfrac{d^{\frac{1}{2}}f}{dx^{\frac{1}{2}}}$	$\dfrac{d^{-\frac{1}{2}}f}{dx^{-\frac{1}{2}}}$
$\beta x \dfrac{b_1,\ldots,b_K}{c_1,\ldots,c_L}$	$\dfrac{1}{\sqrt{x}}\left[\beta x \dfrac{0, b_1,\ldots,b_K}{-\frac{1}{2}, c_1,\ldots,c_L}\right]$	$\sqrt{x}\left[\beta x \dfrac{0, b_1,\ldots,b_K}{\frac{1}{2}, c_1,\ldots,c_L}\right]$
$\beta x \dfrac{-\frac{1}{2}, b_2,\ldots,b_K}{c_1,\ldots,c_L}$	$\dfrac{1}{\sqrt{x}}\left[\beta x \dfrac{0, b_2,\ldots,b_K}{c_1,\ldots,c_L}\right]$	$\sqrt{x}\left[\beta x \dfrac{0, -\frac{1}{2}, b_2,\ldots,b_K}{\frac{1}{2}, c_1,\ldots,c_L}\right]$
$\beta x \dfrac{\frac{1}{2}, b_2,\ldots,b_K}{c_1,\ldots,c_L}$	$\dfrac{1}{\sqrt{x}}\left[\beta x \dfrac{0, \frac{1}{2}, b_2,\ldots,b_K}{-\frac{1}{2}, c_1,\ldots,c_L}\right]$	$\sqrt{x}\left[\beta x \dfrac{0, b_2,\ldots,b_K}{c_1,\ldots,c_L}\right]$
$\beta x \dfrac{b_1,\ldots,b_K}{0, c_2,\ldots,c_L}$	$\dfrac{1}{\sqrt{x}}\left[\beta x \dfrac{b_1,\ldots,b_K}{-\frac{1}{2}, c_2,\ldots,c_L}\right]$	$\sqrt{x}\left[\beta x \dfrac{b_1,\ldots,b_K}{\frac{1}{2}, c_2,\ldots,c_L}\right]$
$\beta x \dfrac{-\frac{1}{2}, b_2,\ldots,b_K}{0, c_2,\ldots,c_L}$	$\dfrac{1}{\sqrt{x}}\left[\beta x \dfrac{b_2,\ldots,b_K}{c_2,\ldots,c_L}\right]$	$\sqrt{x}\left[\beta x \dfrac{-\frac{1}{2}, b_2,\ldots,b_K}{\frac{1}{2}, c_2,\ldots,c_L}\right]$
$\beta x \dfrac{\frac{1}{2}, b_2,\ldots,b_K}{0, c_2,\ldots,c_L}$	$\dfrac{1}{\sqrt{x}}\left[\beta x \dfrac{\frac{1}{2}, b_2,\ldots,b_K}{-\frac{1}{2}, c_2,\ldots,c_L}\right]$	$\sqrt{x}\left[\beta x \dfrac{b_2,\ldots,b_K}{c_2,\ldots,c_L}\right]$
$_p\left[\beta x \dfrac{\ldots}{\ldots}\right]$	$x^{p-\frac{1}{2}}\left[\beta x \dfrac{p, \ldots}{p-\frac{1}{2}, \ldots}\right]$	$x^{p+\frac{1}{2}}\left[\beta x \dfrac{p, \ldots}{p+\frac{1}{2}, \ldots}\right]$
$_p\left[\beta x \dfrac{p-\frac{1}{2}, \ldots}{\ldots}\right]$	$x^{p-\frac{1}{2}}\left[\beta x \dfrac{p, \ldots}{\ldots}\right]$	$x^{p+\frac{1}{2}}\left[\beta x \dfrac{p-\frac{1}{2}, p, \ldots}{p+\frac{1}{2}, \ldots}\right]$
$_p\left[\beta x \dfrac{p+\frac{1}{2}, \ldots}{\ldots}\right]$	$x^{p-\frac{1}{2}}\left[\beta x \dfrac{p, p+\frac{1}{2}, \ldots}{p-\frac{1}{2}, \ldots}\right]$	$x^{p+\frac{1}{2}}\left[\beta x \dfrac{p, \ldots}{\ldots}\right]$
$_p\left[\beta x \dfrac{\ldots}{p, \ldots}\right]$	$x^{p-\frac{1}{2}}\left[\beta x \dfrac{\ldots}{p-\frac{1}{2}, \ldots}\right]$	$x^{p+\frac{1}{2}}\left[\beta x \dfrac{\ldots}{p+\frac{1}{2}, \ldots}\right]$
$_p\left[\beta x \dfrac{p-\frac{1}{2}, \ldots}{p, \ldots}\right]$	$x^{p-\frac{1}{2}}\left[\beta x \dfrac{\ldots}{\ldots}\right]$	$x^{p+\frac{1}{2}}\left[\beta x \dfrac{p-\frac{1}{2}, \ldots}{p+\frac{1}{2}, \ldots}\right]$

table continues

7 SEMIDERIVATIVES AND SEMIINTEGRALS

f	$\dfrac{d^{\frac{1}{2}}f}{dx^{\frac{1}{2}}}$	$\dfrac{d^{-\frac{1}{2}}f}{dx^{-\frac{1}{2}}}$
$x^p\left[\beta x \dfrac{p+\frac{1}{2},\ldots}{p,\ldots}\right]$	$x^{p-\frac{1}{2}}\left[\beta x \dfrac{p+\frac{1}{2},\ldots}{p-\frac{1}{2},\ldots}\right]$	$x^{p+\frac{1}{2}}\left[\beta x \dfrac{\ldots}{\ldots}\right]$
$\left[\beta x^2 \dfrac{\ldots}{\ldots}\right]$	$\sqrt{\dfrac{2}{x}}\left[\beta x^2 \dfrac{-\frac{1}{2},0,\ldots}{-\frac{3}{4},-\frac{1}{4},\ldots}\right]$	$\sqrt{\dfrac{x}{2}}\left[\beta x^2 \dfrac{-\frac{1}{2},0,\ldots}{-\frac{1}{4},\frac{1}{4},\ldots}\right]$
$x^p\left[\beta x^2 \dfrac{\ldots}{\ldots}\right]$	$\sqrt{2}\,x^{p-\frac{1}{2}}\left[\beta x^2 \dfrac{\frac{1}{2}p-\frac{1}{2},\frac{1}{2}p,\ldots}{\frac{1}{2}p-\frac{3}{4},\frac{1}{2}p-\frac{1}{4},\ldots}\right]$	$\dfrac{x^{p+\frac{1}{2}}}{\sqrt{2}}\left[\beta x^2 \dfrac{\frac{1}{2}p-\frac{1}{2},\frac{1}{2}p,\ldots}{\frac{1}{2}p-\frac{1}{4},\frac{1}{2}p+\frac{1}{4},\ldots}\right]$

Notice that semidifferentiation or semiintegration of a $\frac{K}{L}$ hypergeometric of argument x leads generally to a $\frac{K+1}{L+1}$ hypergeometric, but that parametric cancellation may reduce this complexity to $\frac{K}{L}$ or $\frac{K-1}{L-1}$. Likewise semidifferintegration of a $\frac{K}{L}$ hypergeometric of argument x^2 leads in general to a complexity of $\frac{K+2}{L+2}$. Cancellation of parameters in the x^2 case, however, may alter the complexity to $\frac{K+1}{L+1}$, $\frac{K}{L}$, $\frac{K-1}{L-1}$, or even to $\frac{K-2}{L-2}$.

7.9 MISCELLANEOUS FUNCTIONS

The functions collected here are the logarithm, the two complete elliptic integrals, and the Heaviside function.

f	$\dfrac{d^{\frac{1}{2}}f}{dx^{\frac{1}{2}}}$	$\dfrac{d^{-\frac{1}{2}}f}{dx^{-\frac{1}{2}}}$
$\ln(x)$	$\dfrac{\ln(4x)}{\sqrt{\pi x}}$	$2\sqrt{\dfrac{x}{\pi}}\,[\ln(4x)-2]$
$\sqrt{x}\,\ln(x)$	$\dfrac{\sqrt{\pi}}{2}\left[\ln\!\left(\dfrac{x}{4}\right)+2\right]$	$\dfrac{x\sqrt{\pi}}{2}\left[\ln\!\left(\dfrac{x}{4}\right)+1\right]$
$\dfrac{\ln(x)}{\sqrt{x}}$	$\dfrac{\sqrt{\pi}}{x}$	$\sqrt{\pi}\,\ln\!\left(\dfrac{x}{4}\right)$

f	$\dfrac{d^{\frac{1}{2}}f}{dx^{\frac{1}{2}}}$	$\dfrac{d^{-\frac{1}{2}}f}{dx^{-\frac{1}{2}}}$
$K(x)$	$\dfrac{\sqrt{\pi}}{2\sqrt{x[1-x]}}$	$\sqrt{\pi}\arcsin(\sqrt{x})$
$E(x)$	$\dfrac{1}{2}\sqrt{\dfrac{\pi[1-x]}{x}}$	$\dfrac{\sqrt{\pi x[1-x]}}{2}+\dfrac{\sqrt{\pi}\arcsin(\sqrt{x})}{2}$
$H(x-x_0)$	$\dfrac{H(x-x_0)}{\sqrt{\pi[x-x_0]}}$	$\dfrac{2\sqrt{x-x_0}\,H(x-x_0)}{\sqrt{\pi}}$

CHAPTER 8

TECHNIQUES IN THE FRACTIONAL CALCULUS

In this chapter we discuss some of the techniques that have been developed for handling differintegration to noninteger order. Because of the special importance of the $q = \pm\frac{1}{2}$ cases, emphasis is placed on semidifferentiation and semiintegration.

Though most of the chapter deals purely with mathematics, Section 8.3 shows how the operations of the fractional calculus may be carried out using electrical circuitry. The ability to perform these operations by such analog techniques permits a hardware implementation of the fractional calculus (Meyer, 1960; Holub and Nemeč, 1966; Allegre *et al.*, 1970; Ichise *et al.*, 1971; Oldham, 1973a).

8.1 LAPLACE TRANSFORMATION

In this section we seek to Laplace transform $d^q f/dx^q$ for all q and differintegrable f, i.e., we wish to relate

$$\mathscr{L}\left\{\frac{d^q f}{dx^q}\right\} \equiv \int_0^\infty \exp(-sx) \frac{d^q f}{dx^q} dx$$

to the Laplace transform $\mathscr{L}\{f\}$ of the differintegrable function. Let us first recall the well-known transforms of integer-order derivatives

$$\mathscr{L}\left\{\frac{d^q f}{dx^q}\right\} = s^q \mathscr{L}\{f\} - \sum_{k=0}^{q-1} s^{q-1-k} \frac{d^k f}{dx^k}(0), \qquad q = 1, 2, 3, \ldots,$$

and multiple integrals

(8.1.1) $$\mathscr{L}\left\{\frac{d^q f}{dx^q}\right\} = s^q \mathscr{L}\{f\}, \qquad q = 0, -1, -2, \ldots$$

134   8 TECHNIQUES IN THE FRACTIONAL CALCULUS

and note that both formulas are embraced by

(8.1.2) $\quad \mathscr{L}\left\{\dfrac{d^q f}{dx^q}\right\} = s^q \mathscr{L}\{f\} - \displaystyle\sum_{k=0}^{q-1} s^k \dfrac{d^{q-1-k}f}{dx^{q-1-k}}(0), \qquad q = 0, \pm 1, \pm 2, \ldots .$

Notice that the upper summation limit, written as $q - 1$ in (8.1.2), may be replaced by any integer larger than $q - 1$ and even by ∞. The only effect of the replacement is to add terms whose coefficients are $d^{-1}f(0)/dx^{-1}$, $d^{-2}f(0)/dx^{-2}$, etc. Such coefficients are necessarily zero for any function f whose Laplace transform exists. We next show that formula (8.1.2) generalizes to include noninteger q by the simple extension

(8.1.3) $\quad \mathscr{L}\left\{\dfrac{d^q f}{dx^q}\right\} = s^q \mathscr{L}\{f\} - \displaystyle\sum_{k=0}^{n-1} s^k \dfrac{d^{q-1-k}f}{dx^{q-1-k}}(0), \qquad \text{all}\quad q,$

where n is the integer such that $n - 1 < q \leq n$. The sum is empty and vanishes when $q \leq 0$.

In proving (8.1.3), we first consider $q < 0$, so that the Riemann–Liouville definition

$$\dfrac{d^q f}{dx^q} = \dfrac{1}{\Gamma(-q)} \int_0^x \dfrac{f(y)\,dy}{[x-y]^{q+1}}, \qquad q < 0,$$

may be adopted. Direct application of the convolution theorem (Churchill, 1948)

$$\mathscr{L}\left\{\int_0^x f_1(x-y)f_2(y)\,dy\right\} = \mathscr{L}\{f_1\}\mathscr{L}\{f_2\}$$

then gives

(8.1.4) $\quad \mathscr{L}\left\{\dfrac{d^q f}{dx^q}\right\} = \dfrac{1}{\Gamma(-q)} \mathscr{L}\{x^{-1-q}\}\mathscr{L}\{f\} = s^q \mathscr{L}\{f\}, \qquad q < 0,$

so that equation (8.1.1) generalizes unchanged for negative q.

For noninteger positive q, we use the result (3.2.5),

$$\dfrac{d^q f}{dx^q} = \dfrac{d^n}{dx^n} \dfrac{d^{q-n}f}{dx^{q-n}},$$

where n is the integer such that $n - 1 < q < n$. Now, on application of formula (8.1.2), we find

$$\mathscr{L}\left\{\dfrac{d^q f}{dx^q}\right\} = \mathscr{L}\left\{\dfrac{d^n}{dx^n}\left[\dfrac{d^{q-n}f}{dx^{q-n}}\right]\right\} = s^n \mathscr{L}\left\{\dfrac{d^{q-n}f}{dx^{q-n}}\right\} - \displaystyle\sum_{k=0}^{n-1} s^k \dfrac{d^{n-1-k}}{dx^{n-1-k}}\left[\dfrac{d^{q-n}f}{dx^{q-n}}\right](0).$$

The difference $q - n$ being negative, the first right-hand term may be evaluated by use of (8.1.4). Since $q - n < 0$, the composition rule may be applied

8.1 LAPLACE TRANSFORMATION

to the terms within the summation (see Sections 5.7 and 5.1). The result

$$\mathscr{L}\left\{\frac{d^q f}{dx^q}\right\} = s^q \mathscr{L}\{f\} - \sum_{k=0}^{n-1} s^k \frac{d^{q-1-k} f}{dx^{q-1-k}}(0), \qquad 0 < q \neq 1, 2, 3, \ldots,$$

follows from these two operations and is seen to be incorporated in (8.1.3).

The transformation (8.1.3) is a very simple generalization of the classical formula for the Laplace transform of the derivative or integral of f. No similar generalization exists, however, for the classical formulas

$$\mathscr{L}\left\{\frac{-f}{x}\right\} = \frac{d^{-1}\mathscr{L}\{f\}}{ds^{-1}}(s) - \frac{d^{-1}\mathscr{L}\{f\}}{ds^{-1}}(\infty),$$

$$\mathscr{L}\{-xf\} = \frac{d\mathscr{L}\{f\}}{ds},$$

(8.1.5) $\qquad \mathscr{L}\{[-x]^n f\} = \frac{d^n \mathscr{L}\{f\}}{ds^n}, \qquad n = 1, 2, 3, \ldots,$

for the integration or differentiation of the transform. That this is so may be established by considering the example $\mathscr{L}\{f\} = s^p$, where $-1 < p < 0$. Then

$$\frac{d^q \mathscr{L}\{f\}}{[ds]^q} = \frac{\Gamma(p+1) s^{p-q}}{\Gamma(p-q+1)} = \frac{\Gamma(p+1)}{\Gamma(p-q+1)} \mathscr{L}\left\{\frac{x^{q-p-1}}{\Gamma(q-p)}\right\}$$

$$= \frac{\Gamma(p+1)\Gamma(-p)}{\Gamma(p-q+1)\Gamma(q-p)} \mathscr{L}\{x^q f\} = \frac{\csc(\pi[-p]) \mathscr{L}\{x^q f\}}{\csc(\pi[q-p])}$$

$$= [\cos(\pi q) - \cot(\pi p) \sin(\pi q)] \mathscr{L}\{x^q f\}.$$

Notice that only for integer values of q is the coefficient of $\mathscr{L}\{x^q f\}$ independent of the function f, even for such a simple example as $f = x^p$. It is futile, therefore, to seek a simple generalization of equation (8.1.5) to noninteger order.

As a final result of this section we shall establish the useful formula

(8.1.6) $\qquad \mathscr{L}\left\{\exp(-kx) \frac{d^q}{dx^q}[f \exp(kx)]\right\} = [s+k]^q \mathscr{L}\{f\}, \qquad q \leq 0,$

of which equation (8.1.4) may be regarded as the $k = 0$ instance. For $q = 0$, equation (8.1.6) is a trivial identity, while for $q < 0$, we may use the Riemann–Liouville definition of the differintegral to write

$$\mathscr{L}\left\{\exp(-kx) \frac{d^q}{dx^q}[\exp(kx) f(x)]\right\} = \mathscr{L}\left\{\frac{1}{\Gamma(-q)} \int_0^x \frac{\exp(-k[x-y]) \mathscr{L}\{f(y)\} \, dy}{[x-y]^{q+1}}\right\}$$

$$= \frac{1}{\Gamma(-q)} \mathscr{L}\left\{\frac{\exp(-kx)}{x^{q+1}}\right\} \mathscr{L}\{f(x)\}$$

136 8 TECHNIQUES IN THE FRACTIONAL CALCULUS

by the convolution theorem. The proof is completed by making use of

$$\mathscr{L}\left\{\frac{\exp(-kx)}{x^{q+1}}\right\} = \Gamma(-q)[s+k]^q$$

[see Churchill (1948)] for $q < 0$.

8.2 NUMERICAL DIFFERINTEGRATION

In this section, algorithms for effecting differintegration to order q will be devised and evaluated. These algorithms are designed to approximate $d^q f/dx^q$ for arbitrary q when the value of f is known at $N+1$ evenly spaced points in the range 0 to x of the independent variable.[1] The nomenclature

$$f_N \equiv f(0),$$

$$f_{N-1} \equiv f\left(\frac{x}{N}\right),$$

$$\vdots$$

$$f_j \equiv f\left(x - \frac{jx}{N}\right),$$

$$\vdots$$

$$f_0 \equiv f(x)$$

for the function values will be adopted. This possibly confusing notation is clarified by reference to Fig. 8.2.1, which depicts an $N = 5$ example.

What is probably the simplest algorithm is generated from the Grünwald definition (3.2.1) of differintegration

$$\frac{d^q f}{[d(x-a)]^q} = \lim_{N\to\infty}\left\{\frac{\left[\frac{x-a}{N}\right]^{-q}}{\Gamma(-q)}\sum_{j=0}^{N-1}\frac{\Gamma(j-q)}{\Gamma(j+1)}f\left(x - j\left[\frac{x-a}{N}\right]\right)\right\}$$

by simply omitting the $N \to \infty$ operation. Thence we find, setting $a = 0$,

(8.2.1) $$\frac{d^q f}{dx^q} \approx \left(\frac{d^q f}{dx^q}\right)_{G1} = \frac{x^{-q}N^q}{\Gamma(-q)}\sum_{j=0}^{N-1}\frac{\Gamma(j-q)}{\Gamma(j+1)}f\left(x - \frac{jx}{N}\right)$$

$$= \frac{x^{-q}N^q}{\Gamma(-q)}\sum_{j=0}^{N-1}\frac{\Gamma(j-q)}{\Gamma(j+1)}f_j.$$

---

[1] Not all algorithms, however, utilize exactly $N+1$ values of f. The $G1$-algorithm employs only N data, f_N being unused. The $G2$- and $L2$-algorithms utilize $N+2$ points, needing an $f_{-1}[\equiv f([N+1]x/N)]$ datum in addition to the standard $N+1$.

8.2 NUMERICAL DIFFERINTEGRATION

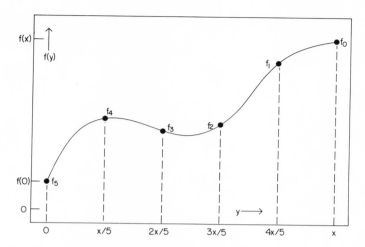

FIG. 8.2.1. An example illustrating the nomenclature used in numerical differintegration.

This is the approximation formula which we term the "$G1$-algorithm." Because of the recursion

$$\frac{\Gamma(j-q)}{\Gamma(j+1)} = \frac{j-1-q}{j} \frac{\Gamma(j-1-q)}{\Gamma(j)},$$

the $G1$-algorithm may be implemented by the convenient multiplication–addition–multiplication $\cdots$ multiplication–addition scheme,

$$(8.2.2) \quad \left(\frac{d^q f}{dx^q}\right)_{G1} = \frac{N^q}{x^q} \Bigg[\Bigg[\cdots \Bigg[\Bigg[f_{N-1}\left\{\frac{N-q-2}{N-1}\right\} + f_{N-2}\Bigg]\left\{\frac{N-q-3}{N-2}\right\}$$

$$+ f_{N-3}\Bigg] \cdots \Bigg]\left\{\frac{1-q}{2}\right\} + f_1\Bigg]\left\{\frac{-q}{1}\right\} + f_0\Bigg],$$

which avoids explicit use of gamma functions and which lends itself to ease of programming.

In Section 3.4 a modified Grünwald definition was presented which was superior in its convergence properties. Rewriting equation (3.4.6) gives

$$(8.2.3) \quad \frac{d^q f}{dx^q} = \lim_{N \to \infty} \left\{ \frac{x^{-q} N^q}{\Gamma(-q)} \sum_{j=0}^{N-1} \frac{\Gamma(j-q)}{\Gamma(j+1)} f\left(x + \frac{qx}{2N} - \frac{jx}{N}\right) \right\}$$

8 TECHNIQUES IN THE FRACTIONAL CALCULUS

for the $a = 0$ case. Unfortunately (unless $q = 0, \pm 2, \pm 4, \ldots$) this formula calls for the evaluation of f at points other than our known f_j values. We therefore use the approximation

$$f\left(x + \frac{qx}{2N} - \frac{jx}{N}\right) \approx \left[\frac{q}{4} + \frac{q^2}{8}\right] f\left(x + \frac{x}{N} - \frac{jx}{N}\right) + \left[1 - \frac{q^2}{4}\right] f\left(x - \frac{jx}{N}\right)$$

$$+ \left[\frac{q^2}{8} - \frac{q}{4}\right] f\left(x - \frac{x}{N} - \frac{jx}{N}\right),$$

based upon the Lagrange three-point interpolation [see Abramowitz and Stegun (1964, p. 879)] in formula (8.2.3) and use this to generate a $G2$-algorithm

$$\left(\frac{d^q f}{dx^q}\right)_{G2} = \frac{x^{-q} N^q}{\Gamma(-q)} \sum_{j=0}^{N-1} \frac{\Gamma(j-q)}{\Gamma(j+1)} \{f_j + \tfrac{1}{4} q[f_{j-1} - f_{j+1}] + \tfrac{1}{8} q^2 [f_{j-1} - 2f_j + f_{j+1}]\}$$

by relaxing the $N \to \infty$ condition. As with the $G1$-algorithm, a multiplicative–additive scheme [akin to (8.2.2)] is a convenient aid to programming this second algorithm.

The two algorithms thus far considered were based on the Grünwald definition of differintegration; the remainder originate with the Riemann–Liouville definition. The latter is

$$\frac{d^q f}{dx^q} = \frac{1}{\Gamma(-q)} \int_0^x \frac{f(y) \, dy}{[x-y]^{q+1}} = \frac{1}{\Gamma(-q)} \int_0^x \frac{f(x-y) \, dy}{y^{q+1}}$$

for negative values of q. Writing this as

$$\frac{d^q f}{dx^q} = \frac{1}{\Gamma(-q)} \sum_{j=0}^{N-1} \int_{jx/N}^{[jx+x]/N} \frac{f(x-y) \, dy}{y^{q+1}},$$

we can see how it is possible to produce a number of so-called R-algorithms, each arising from a different approximation for the integral. Thus, by approximating

$$\int_{jx/N}^{[jx+x]/N} \frac{f(x-y) \, dy}{y^{q+1}} \approx \frac{f\left(x - \frac{jx}{N}\right) + f\left(x - \frac{x}{N} - \frac{jx}{N}\right)}{2} \int_{jx/N}^{[jx+x]/N} \frac{dy}{y^{q+1}}$$

$$= \frac{f_j + f_{j+1}}{-2q} \left\{ \left[\frac{jx+x}{N}\right]^{-q} - \left[\frac{jx}{N}\right]^{-q} \right\},$$

we arrive at the *R*1-algorithm,[2]

(8.2.4) $$\left(\frac{d^q f}{dx^q}\right)_{R1} = \frac{x^{-q} N^q}{\Gamma(1-q)} \sum_{j=0}^{N-1} \frac{f_j + f_{j+1}}{2} \{[j+1]^{-q} - j^{-q}\},$$

valid for $q < 0$. Figure 8.2.2 will explain the basis for the approximation we have just used, as well as demonstrating its inferiority to the approximation

$$\int_{jx/N}^{[jx+x]/N} \frac{f(x-y)\,dy}{y^{q+1}}$$

$$\approx \int_{jx/N}^{[jx+x]/N} \frac{\left\{\left[1+j-\frac{Ny}{x}\right] f\left(x-\frac{jx}{N}\right) + \left[\frac{Ny}{x}-j\right] f\left(x-\frac{x}{N}-\frac{jx}{N}\right)\right\} dy}{y^{q+1}},$$

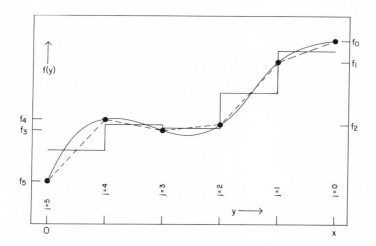

FIG. 8.2.2. The curve *f* is replaced by a piecewise-constant (staircase line) or a piecewise-linear (dashed line) approximation for the purpose of deriving *R*-algorithms.

---

[2] An alternative derivation of the *R*1-algorithm replaces $f(x)$ by the "staircase approximation" shown in Fig. 8.2.2 and differintegrates this approximation by treating it as a piecewise-defined function as in Section 6.9. The *R*2-algorithm likewise arises if the "ramp approximation" of Fig. 8.2.2 is piecewise differintegrated. Note that such techniques of numerical differintegration can equally be applied to data in which the abscissas are not equally spaced.

based on a linear interpolation between f_{j+1} and f_j. This latter leads directly to the R2-algorithm,

(8.2.5) $\left(\dfrac{d^q f}{dx^q}\right)_{R2} = \dfrac{x^{-q} N^q}{\Gamma(-q)} \sum_{j=0}^{N-1} \dfrac{[j+1]f_j - jf_{j+1}}{-q} \{[j+1]^{-q} - j^{-q}\}$

$+ \dfrac{f_{j+1} - f_j}{1 - q} \{[j+1]^{1-q} - j^{1-q}\}$

more complex than the R1, but more precise.

Both R-algorithms are restricted to negative q. To construct an algorithm valid in the range $0 \leq q < 1$ of the differintegration order, we turn to definition (3.6.4). Written for $a = 0$ and $n = 1$, this formula gives

$$\dfrac{d^q f}{dx^q} = \dfrac{x^{-q} f(0)}{\Gamma(1 - q)} + \dfrac{1}{\Gamma(1 - q)} \int_0^x \left[\dfrac{df}{dy}(y)\right] \dfrac{dy}{[x - y]^q}$$

$$= \dfrac{1}{\Gamma(1 - q)} \left\{ \dfrac{f(0)}{x^q} + \sum_{j=0}^{N-1} \int_{jx/N}^{[jx+x]/N} \left[\dfrac{df}{dy}(x - y)\right] \dfrac{dy}{y^q} \right\}.$$

We shall put forward only one algorithm, the L1-algorithm, based on this formula, although more sophisticated alternatives are available. The L1-algorithm utilizes the approximation

$$\int_{jx/N}^{[jx+x]/N} \left[\dfrac{df}{dy}(x - y)\right] \dfrac{dy}{y^q} \approx \dfrac{f\left(x - \dfrac{jx}{N}\right) - f\left(x - \dfrac{x}{N} - \dfrac{jx}{N}\right)}{x/N} \int_{jx/N}^{[jx+x]/N} \dfrac{dy}{y^q}$$

$$= \dfrac{x^{-q} N^q}{1 - q} [f_j - f_{j+1}][(j+1)^{1-q} - j^{1-q}]$$

and is

(8.2.6)

$$\left(\dfrac{d^q f}{dx^q}\right)_{L1} = \dfrac{x^{-q} N^q}{\Gamma(2 - q)} \left[\dfrac{[1 - q]f_N}{N^q} + \sum_{j=0}^{N-1} [f_j - f_{j+1}][(j+1)^{1-q} - j^{1-q}]\right],$$

being valid only for $0 \leq q < 1$.

Turning now to the $1 \leq q < 2$ range, we can once more start with definition (3.6.4), this time choosing $a = 0$ and $n = 2$. The resulting formula,

$$\dfrac{d^q f}{dx^q} = \dfrac{x^{-q} f(0)}{\Gamma(1 - q)} + \dfrac{x^{1-q} f^{(1)}(0)}{\Gamma(2 - q)} + \dfrac{1}{\Gamma(2 - q)} \int_0^x \dfrac{f^{(2)}(y)\, dy}{[x - y]^{q-1}}$$

$$= \dfrac{1}{\Gamma(2 - q)} \left[\dfrac{[1 - q]f(0)}{x^q} + \dfrac{f^{(1)}(0)}{x^{q-1}} + \sum_{j=0}^{N-1} \int_{jx/N}^{[jx+x]/N} \dfrac{f^{(2)}(x - y)\, dy}{y^{q-1}}\right],$$

will serve as our base from which to construct an *L2*-algorithm. By making the approximations

$$f^{(1)}(0) \approx \frac{f\left(\frac{x}{N}\right) - f(0)}{x/N} = \frac{N}{x}[f_{N-1} - f_N]$$

and

$$\int_{jx/N}^{[jx+x]/N} f^{(2)}(x-y)\frac{dy}{y^{q-1}}$$

$$\approx \frac{f\left(x + \frac{x}{N} - \frac{jx}{N}\right) - 2f\left(x - \frac{jx}{N}\right) + f\left(x - \frac{x}{N} - \frac{jx}{N}\right)}{[x/N]^2} \int_{jx/N}^{[jx+x]/N} \frac{dy}{y^{q-1}}$$

$$= \frac{x^{-q}N^q}{2-q}[f_{j-1} - 2f_j + f_{j+1}][(j+1)^{2-q} - j^{2-q}],$$

we arrive at

(8.2.7) $$\left(\frac{d^q f}{dx^q}\right)_{L2} = \frac{x^{-q}N^q}{\Gamma(3-q)}\left[\frac{[1-q][2-q]f_N}{N^q} + \frac{[2-q][f_{N-1} - f_N]}{N^{q-1}}\right.$$
$$\left. + \sum_{j=0}^{N-1}[f_{j-1} - 2f_j + f_{j+1}][(j+1)^{2-q} - j^{2-q}]\right]$$

as our newest algorithm, valid for $1 \leq q < 2$.

It is evident that an *L3*-algorithm covering the $2 \leq q < 3$ range, and an *L4*-algorithm covering the $3 \leq q < 4$ range, etc., may be devised in an analogous fashion. We shall not, however, pursue any of these algorithms further.

In general, algorithms are imperfect. A measure of the imperfection of a typical *Al*-algorithm is given by the relative error term

$$(\varepsilon(f))_{Al} \equiv \frac{\left(\frac{d^q f}{dx^q}\right)_{Al} - \left(\frac{d^q f}{dx^q}\right)}{\frac{d^q f}{dx^q}}$$

for some trial function *f*. This error term will generally depend on q and N,[3] and should tend to zero as N approaches infinity. For certain simple

---

[3] Possibly on x also, though not in the examples considered here.

142   8 TECHNIQUES IN THE FRACTIONAL CALCULUS

Table 8.2.1.   *Comparison of the relative errors of six differintegration algorithms for the functions C, x, and x^2*

Al	$(\varepsilon(C))_{Al}$	$(\varepsilon(x))_{Al}$	$(\varepsilon(x^2))_{Al}$
G1 (all q)	$\dfrac{q[q+1]}{2N}$	$\dfrac{q[q-1]}{2N}$	$\dfrac{q[q-2]}{2N}$
G2 (all q)	$\dfrac{q[q+1]}{2N}$	$\dfrac{q^2}{2N}$	$\dfrac{q[q-1][q-2]}{24N^2}$
R1 ($q<0$)	0	$\dfrac{1-q}{N}\left[\dfrac{\zeta(q)}{N^{-q}} - \dfrac{q}{12N}\right]$	$\dfrac{[2-q][1-q]}{N^2}\left[\dfrac{\zeta(q)}{N^{-q-1}} - \dfrac{\zeta(q-1)}{N^{-q}} + \dfrac{1}{6}\right]$
R2 ($q<0$)	0	0	$\dfrac{2-q}{N^2}\left[\dfrac{\zeta(1-q)}{N^{-q}} + \dfrac{1-q}{12}\right]$
L1 ($0 \leq q < 1$)	0	0	$\dfrac{2-q}{N^2}\left[\dfrac{\zeta(1-q)}{N^{-q}} + \dfrac{1-q}{12}\right]$
L2 ($1 \leq q < 2$)	0	0	$\dfrac{2-q}{N}$

functions f, analytical expressions for $(\varepsilon(f))_{Al}$ may be derived that are excellent asymptotic approximations for large N. Table 8.2.1 presents such expressions for six of our differintegration algorithms, and for the three functions

$$f = C \text{ (a constant)}, \quad f = x, \quad \text{and} \quad f = x^2.$$

Notice that, for these simple functions, several algorithms are error free. In this tabulation, $\zeta(\ )$ denotes the Riemann zeta function (Abramowitz and Stegun, 1964, p. 807), some values of which will be found listed in Table 8.2.2. The methods by which Table 8.2.1 was constructed will be explained by consideration of three examples: $(\varepsilon(C))_{G1}$, $(\varepsilon(x))_{R1}$, and $(\varepsilon(x^2))_{L2}$.

From equation (8.2.1) defining the G1-algorithm, we have

$$\left(\frac{d^q C}{dx^q}\right)_{G1} = \frac{x^{-q} N^q}{\Gamma(-q)} C \sum_{j=0}^{N-1} \frac{\Gamma(j-q)}{\Gamma(j+1)}$$

for $f = C$. Using summation (1.3.17),

$$\left(\frac{d^q C}{dx^q}\right)_{G1} = \frac{C x^{-q} N^q}{\Gamma(1-q)} \frac{\Gamma(N-q)}{\Gamma(N)},$$

whereas the exact differintegral is

$$\frac{d^q C}{dx^q} = \frac{C x^{-q}}{\Gamma(1-q)}$$

Table 8.2.2. *Values of the Riemann zeta function*

q	$\zeta(q)$	q	$\zeta(q)$	q	$\zeta(q)$
-4	0	$-\frac{2}{3}$	-0.15519690	$+1$	∞
$-3\frac{1}{2}$	$+0.00444098$	$-\frac{1}{2}$	-0.20788622	$+1\frac{1}{6}$	$+6.58921554$
-3	$+0.00833333$	$-\frac{1}{3}$	-0.27734305	$+1\frac{1}{3}$	$+3.60093775$
$-2\frac{1}{2}$	$+0.00851692$	$-\frac{1}{6}$	-0.37073766	$+1\frac{1}{2}$	$+2.61237535$
-2	0	0	-0.50000000	$+2$	$+1.64493407$
$-1\frac{1}{2}$	-0.02548520	$+\frac{1}{6}$	-0.68658158	$+2\frac{1}{2}$	$+1.34148726$
$-1\frac{1}{3}$	-0.04006133	$+\frac{1}{3}$	-0.97336025	$+3$	$+1.20205690$
$-1\frac{1}{6}$	-0.05896522	$+\frac{1}{2}$	-1.46035451	$+3\frac{1}{2}$	$+1.12673387$
-1	-0.08333333	$+\frac{2}{3}$	-2.44758074	$+4$	$+1.08232323$
$-\frac{5}{6}$	-0.11469997	$+\frac{5}{6}$	-5.43504324	∞	1

by equation (4.2.1). The error term is thus

$$(\varepsilon(C))_{G1} = \frac{N^q \Gamma(N-q)}{\Gamma(N)} - 1 \sim \frac{q[q+1]}{2N} + O(N^{-2}),$$

where the asymptotic expansion is a consequence of (1.3.13) and is valid for large N.

The derivation of $(\varepsilon(x))_{R1}$ proceeds as follows, starting with equation (8.2.4):

$$\left(\frac{d^q x}{dx^q}\right)_{R1} = \frac{x^{-q} N^q}{\Gamma(1-q)} \sum_{j=0}^{N-1} \frac{[N-j]x + [N-j-1]x}{2N} \{[j+1]^{-q} - j^{-q}\}$$

$$= \frac{x^{1-q} N^{q-1}}{2\Gamma(1-q)} \left[\sum_{j=1}^{N} [2N - 2j + 1]j^{-q} - \sum_{j=0}^{N-1} [2N - 2j - 1]j^{-q}\right]$$

$$= \frac{x^{1-q} N^{q-1}}{2\Gamma(1-q)} \left[N^{-q} + 2\sum_{j=1}^{N-1} j^{-q}\right].$$

To progress further, use is made of the asymptotic expansion

$$\sum_{j=1}^{N} j^{-q} \sim \zeta(q) + \frac{N^{1-q}}{1-q} + \frac{N^{-q}}{2} - \frac{qN^{-q-1}}{12} + O(N^{-q-3})$$

[see Oldham (1970)]. Thence, recalling from Chapter 4 that $x^{1-q}/\Gamma(2-q)$ is the exact differintegral of x, we obtain

$$(\varepsilon(x))_{R1} \sim \frac{[1-q]\zeta(q)}{N^{1-q}} - \frac{q[1-q]}{12N^2} + O(N^{-4}),$$

valid as $N \to \infty$, as the appropriate relative error term. Either of the two leading terms may dominate,[4] depending on the value of q, so both are included in the Table 8.2.1 entry.

Setting $f_j = [N - j]^2 x^2/N^2$ in the equation (8.2.7) defining the $L2$-algorithm, one easily arrives at

$$\left(\frac{d^q x^2}{dx^q}\right)_{L2} = \frac{x^{-q} N^q}{\Gamma(3-q)} \left[\frac{2-q}{N^{q-1}} \left\{\frac{x}{N}\right\}^2 + \sum_{j=0}^{N-1} \left\{\frac{2x^2}{N^2}\right\}\{[j+1]^{2-q} - j^{2-q}\}\right]$$

$$= \frac{2x^{2-q}}{\Gamma(3-q)} \left[\frac{2-q}{2N} + 1\right] = \frac{d^q x^2}{dx^q} \left[\frac{2-q}{2N} + 1\right],$$

from which the tabular entry follows exactly.

All of the differintegration algorithms we have devised are expressible as

(8.2.8) $$\left(\frac{d^q f}{dx^q}\right)_{Al} = \frac{N^q}{x^q} \sum_{j=-1}^{N} w_j(q) f_j,$$

where $w_j(q)$ is a weighting factor whose value depends on j, q, and on the algorithm in question, but not on N or f. For most algorithms, the values of the weighting factors near the two ends of the $-1 \leq j \leq N$ range are atypical and cannot be calculated from the $w_j(q)$ formula applicable in the middle of the j range. Table 8.2.3 (pp. 146–147) is comprehensive, listing both typical and atypical weighting factors for all six algorithms. The use of summation (8.2.8) in conjunction with Table 8.2.3 provides an alternative computational method, replacing formulas such as (8.2.1), (8.2.2), (8.2.4), (8.2.5), etc.

Notice in Table 8.2.3 that the same entries serve the $R2$-algorithm and the $L1$-algorithm. In fact, despite the apparent dissimilarity of definitions (8.2.5) and (8.2.6), these two algorithms are identical. Henceforth we shall use the term "RL-algorithm" to refer to this conjunction.

At this point it is appropriate to review the advantages and disadvantages of the various algorithms. The G-algorithms are valuable in that they span all q values, can be implemented by a convenient multiplication-addition scheme, and need no gamma functions. The $G1$-algorithm, moreover, is unique in not utilizing an f_N datum, permitting this algorithm to be used to differintegrate functions akin to $\ln(x)$ and $1/\sqrt{x}$, which are infinite at $x = 0$. The $R1$-algorithm was devised from a formula restricted to negative q; it is inferior to the RL-algorithm in accuracy, but is significantly simpler. The RL-algorithm is an efficient and relatively simple algorithm; it is, in fact, the one which we have adopted in routine work (Grenness and Oldham, 1972; Oldham, 1972). The $L2$-algorithm, designed for the $1 \leq q < 2$ range,

---

[4] Because $\zeta(-1) = -\frac{1}{12}$ (see Table 8.2.2), the error inherent in the $R1$-algorithm is minute for q values in the vicinity of -1.

is complex to implement and (in common with the $G2$-algorithm) requires an f_{-1} datum, a requirement which cannot be met in some physical applications.

The identity of the $R2$- and $L1$-algorithms illustrates the fact that an algorithm, although designed to differintegrate for a limited q-range, may perform efficiently over an extended range. To test this possibility, and to provide a crucial comparative test for our five algorithms, Fig. 8.2.3 and 8.2.4 were constructed.

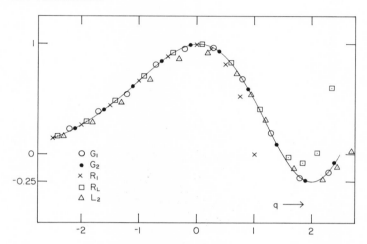

FIG. 8.2.3. The curve depicts the exact differintegral of $\sqrt{x}$, evaluated at $x = 1$. The points are approximations generated by the $G1$, $G2$, $R1$, RL, and $L2$ algorithms, utilizing 32 data points.

The curve in Fig. 8.2.3 is a plot of $\Gamma(\tfrac{3}{2})/\Gamma(\tfrac{3}{2} - q)$ versus q, for a wide range of q values. Each set of points in this diagram was produced by allowing one of the five algorithms to differintegrate $\sqrt{x}$ to order q, the differintegral then being evaluated at $x = 1$. Thus, since

$$\frac{d^q \sqrt{x}}{dx^q}(1) = \frac{\Gamma(\tfrac{3}{2})}{\Gamma(\tfrac{3}{2} - q)},$$

all points would lie on the curve if the algorithms were exact. In fact, since we chose $N = 32$ only, some points are seen to diverge significantly from the curve, particularly at large positive q values.

Figure 8.2.4 is likewise a graph versus q of the function

$$\frac{d^q[1 - x^{\tfrac{3}{2}}]}{dx^q}(1) = \frac{\Gamma(1)}{\Gamma(1 - q)} - \frac{\Gamma(\tfrac{5}{2})}{\Gamma(\tfrac{5}{2} - q)},$$

Table 8.2.3. *Weighting factors for six numerical differintegration algorithms*

Al	Typical $w_j(q)$	Range of typicality	Values of j for which $w_j(q) = 0$	Atypical values
G1	$\dfrac{\Gamma(j-q)}{\Gamma(-q)\Gamma(j+1)}$	$0 \leq j \leq N-1$	$-1, N$	none
G2	$\dfrac{\Gamma(j-1-q)}{\Gamma(-q)\Gamma(j+2)}\left\{j^2 - \dfrac{jq}{2}[q+3] - [q+1]\left[\dfrac{q^3}{8} + \dfrac{q^2}{2} - 1\right]\right\}$	$0 \leq j \leq N-1$	none	$w_{-1}(q) = \dfrac{q^2+2q}{8\Gamma(-q)}$ $w_N(q) = \dfrac{\Gamma(N-1-q)}{\Gamma(-q)\Gamma(N)}\left[\dfrac{q^2}{8} - \dfrac{q}{4}\right]$
R1	$\dfrac{[j+1]^{-q} - [j-1]^{-q}}{\Gamma(1-q)}$	$1 \leq j \leq N-1$	-1	$w_0(q) = \dfrac{1}{2\Gamma(1-q)}$ $w_N(q) = \dfrac{N^{-q} - [N-1]^{-q}}{2\Gamma(1-q)}$

8.2 NUMERICAL DIFFERINTEGRATION

R2	$\dfrac{[j+1]^{1-q} - 2j^{1-q} + [j-1]^{1-q}}{\Gamma(2-q)}$	$1 \leq j \leq N-1$	-1	$w_0(q) = 1/\Gamma(2-q)$ $w_N(q) = \dfrac{[N-1]^{1-q} - N^{1-q} + [1-q]N^{-q}}{\Gamma(2-q)}$
L1	$\dfrac{[j+1]^{1-q} - 2j^{1-q} + [j-1]^{1-q}}{\Gamma(2-q)}$	$1 \leq j \leq N-1$	-1	$w_0(q) = 1/\Gamma(2-q)$ $w_N(q) = \dfrac{[N-1]^{1-q} - N^{1-q} + [1-q]N^{-q}}{\Gamma(2-q)}$
L2	$\dfrac{[j+2]^{2-q} - 3[j+1]^{2-q} + 3j^{2-q} - [j-1]^{2-q}}{\Gamma(3-q)}$	$1 \leq j \leq N-2$	none	$w_{-1}(q) = 1/\Gamma(3-q)$ $w_0(q) = \dfrac{2^{2-q} - 3}{\Gamma(3-q)}$ $w_{N-1}(q) = [\Gamma(3-q)]^{-1}\{[2-q]N^{1-q} - 2N^{2-q} + 3[N-1]^{2-q} - [N-2]^{2-q}\}$ $w_N(q) = [\Gamma(3-q)]^{-1}\{[1-q][2-q]N^{-q} - [2-q]N^{1-q} + N^{2-q} - [N-1]^{2-q}\}$

constructed to display the extent to which each algorithm reproduces the true differintegral, using only 32 data points. The functions $\sqrt{x}$ and $1 - x^{\frac{3}{2}}$ were chosen for these comparisons because, in the range $0 < x \leqq 1$, they place different emphases on the two ends of the range, the former stressing the points close to $x = 1$, the latter emphasizing data close to the $x = 0$ origin.

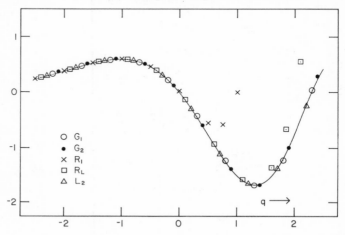

FIG. 8.2.4. The exact differintegral $d^q[1 - x^{\frac{3}{2}}]/dx^q$ at $x = 1$ and approximations to it given by the five algorithms, calculated using $N = 32$.

Inspection of the two diagrams discloses that all algorithms except the $L2$ are efficient for negative q, and extend into the positive q range with varying success. Of course, increasing N beyond 32 will increase the range of efficiency of all the algorithms. The $L2$-algorithm, as befits its mode of construction, is most successful for modestly positive q values.

8.3 ANALOG DIFFERINTEGRATION

Electrical circuits have long been used to perform the operations of differentiation and integration. Probably the simplest example is that shown in Fig. 8.3.1 in which a capacitor is used to provide a voltage output proportional to the integral of the applied current

$$e(t) = \frac{1}{C} \frac{d^{-1}}{dt^{-1}} i(t),$$

provided the initial voltage $e(0)$ is zero. Using the example of semiintegration, this section will demonstrate how circuits can be designed that perform differintegration with respect to time, to noninteger orders.

8.3 ANALOG DIFFERINTEGRATION 149

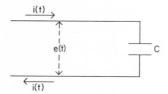

FIG. 8.3.1. An integrating circuit. If $e(0)$ is zero, the potential $e(t)$ across C is proportional to the integral of the input current $i(t)$.

The semiintegrating circuits (for which the name "semiintegrators" seems appropriate) that we shall discuss are imperfect in two respects. First, they are imprecise in that the response is only approximately proportional to the semiintegral of the input signal. Nevertheless, the output may be made to approach the true semiintegral to any specified degree of accuracy. Second, the semiintegration is accomplished only over a time interval which has a finite upper limit and a nonzero lower limit.[5] Nevertheless, these time limits may be chosen to embrace as long and as short a time as desired. Increasing perfection in both accuracy and time span is bought at the price of an increase in the number of components of which the circuit is constructed. Thus, it is a relatively simple matter to design a semiintegrator that performs within an accuracy[6] of $\pm 10\%$ over a time span from 0.1 to 10 sec, whereas a $\pm 1\%$ semiintegrator spanning the time period 10^{-2} to 10^2 sec requires very many more (and closer tolerance) components.

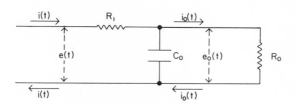

FIG. 8.3.2. A simple three-component circuit.

First consider the simple circuit depicted in Fig. 8.3.2, in which the current signal $i(t)$ must flow through the resistor R_1, following which it experiences a parallel combination of the capacitor C_0 and the resistor R_0. Imagine that

---

[5] We may replace this sentence in the somewhat more exact parlance of communications theory by saying that the semiintegrator has a frequency bandpass.

[6] What exactly is meant by "within an accuracy" will be apparent later. The accuracy criterion is actually prescribed in Laplace transform space, rather than in a real time domain.

prior to $t = 0$ the circuit is at rest, the applied current and the potential output both being zero:

(8.3.1) $$i(t \leq 0) = 0 = e(t \leq 0).$$

A nonzero signal is applied after $t = 0$.

The meanings of the symbols $i_0(t)$ and $e_0(t)$ will be clarified on inspection of Fig. 8.3.2. The current $i_0(t)$ flows through resistor R_0 and therefore is, by Ohm's law, proportional to the potential across the resistor, the constant of proportionality being the reciprocal of the resistance; that is,

$$i_0(t) = \frac{e_0(t)}{R_0}.$$

On the other hand, the current through the capacitor is proportional to the time derivative of the potential across it, the proportionality constant being the capacitance of the capacitor; that is,

$$i(t) - i_0(t) = C_0 \frac{de_0}{dt}(t).$$

Ohm's law also permits us to write

$$i(t) = \frac{e(t) - e_0(t)}{R_1}$$

for the leftmost resistor. The variables $e_0(t)$ and $i_0(t)$ can be eliminated from the above three expressions to yield the differential equation

(8.3.2) $$[R_0 + R_1]i(t) + R_0 R_1 C_0 \frac{di}{dt}(t) = e(t) + R_0 C_0 \frac{de}{dt}(t),$$

interrelating the current $i(t)$ and the potential $e(t)$.

Circuit analysis is frequently more tractable in transform space, and we now therefore consider the Laplace transform,

$$\bar{i}(s)[R_0 + R_1 + R_0 R_1 C_0 s] - R_0 R_1 C_0 i(0) = \bar{e}(s)[1 + R_0 C_0 s] - R_0 C_0 e(0),$$

of equation (8.3.2). Here $\bar{i}(s)$ and $\bar{e}(s)$ are, respectively, the transforms of $i(t)$ and $e(t)$, s being the dummy variable of Laplace transformation with respect to time [see Churchill (1948) for a thorough treatment of Laplace transformation theory and techniques]. This transform equation can be recast as

(8.3.3) $$\frac{\bar{e}(s)}{R_1 \bar{i}(s)} = 1 + \frac{\frac{1}{R_1 C_0}}{s + \frac{1}{R_0 C_0}}$$

after invoking the initial conditions (8.3.1).

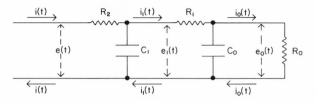

FIG. 8.3.3. Two additional components, a series resistor R_2 and a shunt capacitor C_1, have been added to the circuit in Fig. 8.3.2.

Next, transfer consideration to circuit shown in Fig. 8.3.3, which contains two components, C_1 and R_2, additional to those shown in Fig. 8.3.2. The equations

$$i(t) = \frac{e(t) - e_1(t)}{R_2} \quad \text{and} \quad i(t) - i_1(t) = C_1 \frac{de_1}{dt}(t)$$

interrelate the currents and voltages in the left-hand side of this circuit, and yield

(8.3.4) $$\frac{\bar{e}(s)}{R_2 \bar{i}(s)} = 1 + \frac{\dfrac{1}{R_2 C_1}}{s + \dfrac{\bar{i}_1(s)}{C_1 \bar{e}_1(s)}}$$

on Laplace transformation, combination, and rearrangement. The right-hand side of the circuit, moreover, obeys the transform equation

(8.3.5) $$\frac{\bar{e}_1(s)}{R_1 \bar{i}_1(s)} = 1 + \frac{\dfrac{1}{R_1 C_0}}{s + \dfrac{1}{R_0 C_0}}$$

by analogy with equation (8.3.3).

Progress is aided if we define the frequencies

$$\omega_0 \equiv \frac{1}{R_0 C_0}, \quad \omega_1 \equiv \frac{1}{R_1 C_0}, \quad \omega_2 \equiv \frac{1}{R_1 C_1}, \quad \omega_3 \equiv \frac{1}{R_2 C_1},$$

characteristic of pairs of adjacent components. Using these abbreviations, equations (8.3.4) and (8.3.5) can be combined to give

$$\frac{\bar{e}(s)}{R_2 \bar{i}(s)} = 1 + \frac{\omega_3}{s + \dfrac{\omega_2}{1 + \dfrac{\omega_1}{s + \omega_0}}},$$

an expression that may be rewritten

(8.3.6) $$\frac{\bar{e}(s)}{R_2 \bar{i}(s)} = 1 + \frac{\omega_3}{s+} \frac{\omega_2}{1+} \frac{\omega_1}{s+} \frac{\omega_0}{1}$$

in the notation of continued fractions.

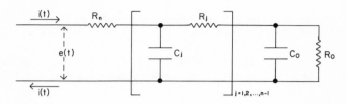

FIG. 8.3.4. A circuit that can perform the operation of semiintegration.

The generalization of equation (8.3.6) to cover the circuit shown in Fig. 8.3.4 is straightforward and yields

(8.3.7) $$\frac{\bar{e}(s)}{R_n \bar{i}(s)} = 1 + \frac{\omega_{2n-1}}{s+} \frac{\omega_{2n-2}}{1+} \frac{\omega_{2n-3}}{s+} \cdots \frac{\omega_2}{1+} \frac{\omega_1}{s+} \frac{\omega_0}{1},$$

where

$$\omega_{2j} = \frac{1}{R_j C_j} \quad \text{and} \quad \omega_{2j+1} = \frac{1}{R_{j+1} C_j}.$$

A simplification ensues if each ω_j is undimensionalized through division by s. Thus if

$$\frac{\omega_j}{s} \equiv v_j,$$

equation (8.3.7) may be expressed as the continued fraction expression

(8.3.8) $$\frac{\bar{e}(s)}{R_n \bar{i}(s)} = 1 + \frac{v_{2n-1}}{1+} \frac{v_{2n-2}}{1+} \frac{v_{2n-3}}{1+} \cdots \frac{v_2}{1+} \frac{v_1}{1+} \frac{v_0}{1}.$$

Only for certain simple relationships between the v values is the continued fraction in equation (8.3.8) expressible in a more convenient form. We shall discuss only one such simplifying circumstance. This arises when all capacitors have a common capacitance value $C_0 = C_1 = C_2 = \cdots = C_{n-1} = C$ and all resistors *except one* have a common resistance. The exceptional resistor is R_n and this is accorded a resistance value equal to one-half that of the others; thus if

$$R_0 = R_1 = R_2 = \cdots = R_{n-1} = R,$$

then

$$R_n = \tfrac{1}{2} R.$$

8.3 ANALOG DIFFERINTEGRATION

If we suitably define v, thus,

$$v_0 = v_1 = v_2 = \cdots = v_{2n-3} = v_{2n-2} = \frac{1}{RCs} \equiv v,$$

then

$$v_{2n-1} = \frac{2}{RCs} = 2v$$

and consequently, from equation (8.3.8),

(8.3.9) $$\frac{2\bar{e}(s)}{R\bar{\imath}(s)} = 1 + 2\frac{v}{1+}\frac{v}{1+}\cdots\frac{v}{1+}\frac{v}{1},$$

where the continued fraction has $2n$ numeratorial v's. Continued fractions are discussed by Wall (1948) who shows that

(8.3.10) $$\frac{v}{1+}\frac{v}{1+}\cdots\frac{v}{1+}\frac{v}{1} = \frac{\sqrt{4v+1}}{1 + \left[\dfrac{\sqrt{4v+1}-1}{\sqrt{4v+1}+1}\right]^{2n+1}} - \frac{\sqrt{4v+1}}{2} - \frac{1}{2},$$

as can also be established inductively. Combining equations (8.3.9) and (8.3.10) and dividing by $2\sqrt{v}$, we obtain

(8.3.11) $$\frac{\bar{e}(s)}{\bar{\imath}(s)}\sqrt{\frac{Cs}{R}} = \sqrt{\frac{4v+1}{4v}} \left[\frac{[\sqrt{4v+1}+1]^{2n+1} - [\sqrt{4v+1}-1]^{2n+1}}{[\sqrt{4v+1}+1]^{2n+1} + [\sqrt{4v+1}-1]^{2n+1}}\right]$$

as our final result interrelating the transforms of $e(t)$ and $i(t)$.

Figure 8.3.5 is a plot of the right-hand side of equation (8.3.11) versus v for various values of n. Notice that, for the larger n values, the function

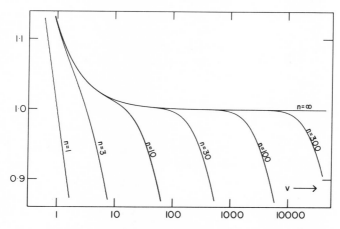

FIG. 8.3.5. A semilogarithmic plot illustrating equation (8.3.11).

is very close to unity over a wide range of v values. In fact, for n in excess of about 10, the function lies within 2% of unity,

$$\frac{\bar{e}(s)}{\bar{i}(s)}\sqrt{\frac{Cs}{R}} \approx 1,$$

provided that

$$6 \leq v \leq \tfrac{1}{6}n^2.$$

Recalling the definition of v, this means that

(8.3.12) $\qquad \bar{e}(s) \approx \sqrt{\dfrac{R}{Cs}}\,\bar{i}(s) \qquad$ for $\qquad 6RC \leq \dfrac{1}{s} \leq \tfrac{1}{6}n^2 RC.$

The Laplace inversion of this relationship poses problems if it is to be performed rigorously; however, (8.3.12) implies

(8.3.13) $\qquad e(t) \approx \sqrt{\dfrac{R}{C}\dfrac{d^{-\frac{1}{2}}i}{dt^{-\frac{1}{2}}}(t)} \qquad$ for $\qquad 6RC \leq t \leq \tfrac{1}{6}n^2 RC$

at least approximately. Thus, over a certain time range, the circuit of Fig. 8.3.4 performs as an efficient semiintegrator in that it develops a potential accurately proportional to the semiintegral of the input current. The lower time limit for accurate semiintegration is about six times the RC time constant of the standard components used, and can be made arbitrarily small by selecting small-valued resistors and capacitors. The upper time limit depends on the number of components and can be made arbitrarily large by employing enough resistors and capacitors.

Although it is mathematically straightforward, the circuitry of Fig. 8.3.4 is not the most efficient semiintegrator. By this we mean that a given number $2n + 1$ of components can be used to construct a semiintegrator that accurately semiintegrates over a wider time span ratio than the $n^2/36$ given by (8.3.13). For details of such circuits the interested reader is referred to Biorci and Ridella (1970), Ichise et al. (1971), and Oldham (1973a).

8.4 EXTRAORDINARY DIFFERENTIAL EQUATIONS

A relationship involving one or more derivatives of an unknown function f with respect to its independent variable x is known as an ordinary differential equation. Similar relationships involving at least one differintegral of noninteger order may be termed extraordinary differential equations. Such an equation is solved when an explicit expression for f is exhibited. As with ordinary differential equations, the solutions of extraordinary differential equations often involve integrals and contain arbitrary constants.

8.4 EXTRAORDINARY DIFFERENTIAL EQUATIONS

We begin by considering perhaps the simplest of all extraordinary differential equations,

(8.4.1) $$\frac{d^Q f}{dx^Q} = F,$$

where Q is arbitrary, F is a known function, and f is the unknown function (we have chosen the lower limit $a = 0$ in the differintegral for simplicity). It is tempting to apply the operator d^{-Q}/dx^{-Q} to both sides of equation (8.4.1) and perform the "inversion"

$$f = \frac{d^{-Q} F}{dx^{-Q}},$$

but this is not the most general solution. In fact, referring to our discussion of the composition law in Section 5.7, we recall that it is precisely the condition

$$f - \frac{d^{-Q}}{dx^{-Q}} \frac{d^Q f}{dx^Q} = 0,$$

which guarantees obedience to the composition rule for general differintegrable series f. The difference $f - d^{-Q}/dx^{-Q}\{d^Q f/dx^Q\}$ will not, in general, vanish but will consist of those portions of the differintegrable series units f_U in f that are sent to zero under the action of d^Q/dx^Q. We decompose f into differintegrable units $f_{U,i}$, where

(8.4.2) $$f_{U,i} \equiv x^{p_i} \sum_{j=0}^{\infty} a_{ij} x^j, \qquad p_i > -1, \quad a_{i0} \neq 0, \quad i = 1, \ldots, n,$$

and investigate the conditions on $f_{U,i}$ required to give

(8.4.3) $$f - \frac{d^{-Q}}{dx^{-Q}} \frac{d^Q f}{dx^Q} \neq 0.$$

Inspection of condition (5.7.5) or of (5.7.6), to which it is equivalent, tells us that condition (8.4.3) obtains if and only if, for some i in the range $1 \leq i \leq n$,

(8.4.4) $$\Gamma(p_i - Q + 1)$$

is infinite. This condition can occur, however, only when $p_i - Q + 1 = 0, -1, -2, \ldots$, that is, when $p_i = Q - 1, Q - 2, \ldots$. Putting these facts together shows that, in the most general case,

$$f - \frac{d^{-Q}}{dx^{-Q}} \frac{d^Q f}{dx^Q} = C_1 x^{Q-1} + C_2 x^{Q-2} + \cdots + C_m x^{Q-m},$$

where $C_1, \ldots, C_m$ are arbitrary constants and $0 < Q \leq m < Q + 1$; for $Q \leq 0$ the right-hand member of the equation is zero. Thus

$$f - C_1 x^{Q-1} - C_2 x^{Q-2} - \cdots - C_m x^{Q-m} = \frac{d^{-Q}}{dx^{-Q}} \frac{d^Q f}{dx^Q} = \frac{d^{-Q} F}{dx^{-Q}}$$

and the most general solution of equation (8.4.1) is

(8.4.5) $$f = \frac{d^{-Q} F}{dx^{-Q}} + C_1 x^{Q-1} + C_2 x^{Q-2} + \cdots + C_m x^{Q-m}.$$

Notice the presence in this general solution of m arbitrary constants, where $0 < Q \leq m < Q + 1$ or $m = 0$ for $Q \leq 0$.

As an example of the foregoing theory, consider the extraordinary differential equation

$$\frac{d^{\frac{3}{2}} f}{dx^{\frac{3}{2}}} = x^5,$$

whose general solution is

$$f = \frac{d^{-\frac{3}{2}} x^5}{dx^{-\frac{3}{2}}} + C_1 x^{\frac{1}{2}} + C_2 x^{-\frac{1}{2}} = \frac{\Gamma(6)}{\Gamma(\frac{15}{2})} x^{13/2} + C_1 x^{\frac{1}{2}} + C_2 x^{-\frac{1}{2}},$$

and contains two arbitrary constants.

Next consider the equation

(8.4.6) $$\frac{d^Q f}{dx^Q} + A \frac{d^{Q-1} f}{dx^{Q-1}} = F(x),$$

where Q is again arbitrary, A is a known constant, and F is known. Application of the operator d^{1-Q}/dx^{1-Q} to equation (8.4.6) yields, by techniques like those discussed in connection with the inversion of equation (8.4.1),

$$\frac{df}{dx} + Af = \frac{d^{1-Q} F}{dx^{1-Q}} + C_1 x^{Q-2} + C_2 x^{Q-3} + \cdots + C_m x^{Q-m-1},$$

a first-order ordinary differential equation for f whose solution may be accomplished by standard methods (Murphy, 1960).

The two extraordinary differential equations just treated were quite special; the solution of the even slightly more general equation

$$\frac{d^q f}{dx^q} + \frac{d^Q f}{dx^Q} = F$$

encounters very great difficulties except when the difference $q - Q$ is integer or half-integer. We have illustrated in the present section how to deal with the case $q - Q = n$, an integer (which leads to an nth-order ordinary differential equation). The next section will treat the important case $q - Q =$

$[2n + 1]/2$, n integer. As we shall see in Chapter 11, these so-called semidifferential equations arise very naturally in solving a wide variety of transport problems and, in our experience, constitute by far the most important class of extraordinary differential equations.

8.5 SEMIDIFFERENTIAL EQUATIONS

By a semidifferential equation we shall understand a relationship involving differintegrals of an unknown function, each differintegral order occurring as some multiple of $\frac{1}{2}$, at least one of which must be an odd multiple of $\frac{1}{2}$. For example, the equation

$$\frac{d^6 f}{dx^6} + \sin(x)\frac{d^{7/2} f}{dx^{7/2}} = \exp(x)$$

is a semidifferential equation, as is

$$\frac{d^{3/2} f}{dx^{3/2}} - \frac{d^{-1/2} f}{dx^{-1/2}} + 2f = 0,$$

while

$$\frac{d^2 f}{dx^2} + \frac{df}{dx} = \frac{d^{1/2} F}{dx^{1/2}}$$

is not if F is regarded as a known function. We shall discover by examples that two principal techniques are available for solving semidifferential equations: (1) Transformation to an ordinary differential equation and (2) Laplace transformation. As is often the case when dealing with the fractional calculus we are not able to discuss solutions of very general semidifferential equations but are forced to content ourselves with examples intended to reveal solution techniques.

Consider the semidifferential equation

(8.5.1) $$\frac{d^{1/2} f}{dx^{1/2}} + f = 0.$$

Applying $d^{1/2}/dx^{1/2}$ and utilizing composition rule arguments as in Section 8.4 yields

$$\frac{df}{dx} - C_1 x^{-3/2} + \frac{d^{1/2} f}{dx^{1/2}} = 0.$$

This result may now be combined with the original semidifferential equation (8.5.1) to give

$$\frac{df}{dx} - f = C_1 x^{-3/2},$$

a first-order ordinary differential equation for the unknown f. Standard methods [see Murphy (1960)] enable us to solve for f, with the result

(8.5.2) $$f = C\exp(x) + \exp(x)\int_0^x \exp(-y)C_1 y^{-3/2}\,dy.$$

The solution (8.5.2) is puzzling on two counts. First, it involves two constants, C_1 and C, rather than the single arbitrary constant we might have expected based on analogy with the results of Section 8.4. Second, it involves a divergent integral, namely,

$$I \equiv \int_0^x \exp(-y) y^{-3/2}\,dy,$$

which we recognize from equation (1.3.26) as the incomplete gamma function $-2\sqrt{\pi/x}\,\gamma^*(-\tfrac{1}{2}, x)$. A finite value may be assigned to I,

$$I = -2\sqrt{\pi}\,\mathrm{erf}(\sqrt{x}) - 2\,\frac{\exp(-x)}{\sqrt{x}},$$

by means of formulas (1.3.27) and (1.3.28). We see then that

(8.5.3) $$f = C\exp(x) - 2C_1 \exp(x)\left[\sqrt{\pi}\,\mathrm{erf}(\sqrt{x}) + \frac{\exp(-x)}{\sqrt{x}}\right]$$

$$= C\exp(x) - 2\sqrt{\pi}\,C_1 \exp(x)\,\mathrm{erf}(\sqrt{x}) - \frac{2C_1}{\sqrt{x}}.$$

Now upon semidifferentiation of equation (8.5.3), making use of Sections 7.5 and 7.3, we see that

$$\frac{d^{1/2}f}{dx^{1/2}} = \frac{C}{\sqrt{\pi x}} + C\exp(x)\,\mathrm{erf}(\sqrt{x}) - 2\sqrt{\pi}\,C_1 \exp(x).$$

The original semidifferential equation now demands that

$$\frac{C}{\sqrt{\pi x}} + C\exp(x)\,\mathrm{erf}(\sqrt{x}) - 2\sqrt{\pi}\,C_1 \exp(x)$$

$$= \frac{2C_1}{\sqrt{x}} + 2\sqrt{\pi}\,C_1 \exp(x)\,\mathrm{erf}(\sqrt{x}) - C\exp(x).$$

This condition is satisfied only if

$$\frac{C}{\sqrt{\pi}} = 2C_1$$

and shows that, indeed, there is but a single arbitrary constant in the solution of (8.5.1), which takes the final form

(8.5.4) $$f = C\exp(x)\,\mathrm{erfc}(\sqrt{x}) - [C/\sqrt{\pi x}\,].$$

Our second example in this section is the semidifferential equation

(8.5.5) $$\frac{df}{dx} + \frac{d^{\frac{1}{2}}f}{dx^{\frac{1}{2}}} - 2f = 0.$$

This time it is useful to Laplace transform, obtaining

$$s\bar{f}(s) - f(0) + \sqrt{s}\,\bar{f}(s) - \frac{d^{-\frac{1}{2}}f}{dx^{-\frac{1}{2}}}(0) + \bar{f}(s) = 0$$

after making use of the results of Section 8.1. Hence

$$\bar{f}(s) = \frac{f(0) + \dfrac{d^{-\frac{1}{2}}f}{dx^{-\frac{1}{2}}}(0)}{s + \sqrt{s} - 2} = \frac{C}{[\sqrt{s} - 1][\sqrt{s} + 2]},$$

where C is a constant. A partial fraction decomposition gives

$$\bar{f}(s) = \frac{C}{3[\sqrt{s} - 1]} - \frac{C}{3[\sqrt{s} + 2]},$$

which upon Laplace inversion produces

$$f = \frac{C}{3}\left[\frac{1}{\sqrt{\pi x}} + \exp(x)\,\mathrm{erfc}(-\sqrt{x})\right] - \frac{C}{3}\left[\frac{1}{\sqrt{\pi x}} - 2\exp(4x)\,\mathrm{erfc}(2\sqrt{x})\right]$$

$$= \frac{C}{3}\left[2\exp(4x)\,\mathrm{erfc}(2\sqrt{x}) + \exp(x)\,\mathrm{erfc}(-\sqrt{x})\right]$$

as the final solution.

The reader has, no doubt, developed a healthy respect for the complexity hidden in solving even the simplest semidifferential equations. In Chapter 11 we reduce the solution of certain diffusion problems "*simply* to that of a semidifferential equation." It is indeed fortunate that such equations usually have the form (8.4.1) with $Q = \pm\frac{1}{2}$, a form which is amenable to easy inversion as demonstrated in Section 8.4.

8.6 SERIES SOLUTIONS

In the same way that resort must often be made to series methods in solving the more difficult ordinary differential equations, so such methods can be used for difficult semidifferential equations, and more generally for extraordinary differential equations. The techniques closely parallel those used for classical differential equations [see Murphy (1960)] and it will suffice here to present a single example.

The semidifferential equation

(8.6.1) $$\sqrt{x}\,\frac{d^{\frac{1}{2}}f}{dx^{\frac{1}{2}}} + x^w f = 1$$

occurs in the theory of voltammetry at expanding electrodes [see Oldham (1969b)]. Here w is a number in the range $0 < w \leq \frac{1}{2}$, the value $w = \frac{3}{14}$ having special practical importance. As encountered practically, w will always be rational, and we shall equate it to the ratio μ/ν.

To solve equation (8.6.1), we assume that the differintegrable series

(8.6.2) $$f = x^p \sum_{j=0}^{\infty} a_j x^{j/n}$$

is a solution for some value of p exceeding -1, for some integer n, and for some assignment of coefficients $a_0, a_1, a_2, \ldots$ such that $a_0 \neq 0$. If so, then we have

$$x^p \sum_{j=0}^{\infty} a_j x^{j/n} \left[\frac{\Gamma(p + \frac{j}{n} + 1)}{\Gamma(p + \frac{j}{n} + \frac{1}{2})} + x^{\mu/\nu} \right] = 1$$

on combining equations (8.6.1) and (8.6.2) and making use of the results of Sections 5.2 and 7.3. To satisfy this equation, we select $p = 0$, $n = \nu$ and the following coefficients:

$$a_0 = \sqrt{\pi}, \quad a_1 = a_2 = \cdots = a_{\mu-1} = 0,$$

$$a_j = -a_{j-\mu} \frac{\Gamma(\frac{j}{\nu} + \frac{1}{2})}{\Gamma(\frac{j}{\nu} + 1)}, \quad j = \mu, \mu+1, \mu+2, \ldots.$$

The semidifferential equation is now effectively solved.

This solution, most concisely written as

$$f_w(x) = \sum_{k=0}^{\infty} [-x]^{kw} \prod_{l=0}^{k} \frac{\Gamma(\frac{1}{2} + lw)}{\Gamma(1 + lw)},$$

was derived independently by Koutecky (1953) and by Matsuda and Ayabe (1955), although not in the context of the fractional calculus and only in the $w = \frac{3}{14}$ instance. Convergence is rapid for small argument, but the asymptotic alternative

$$f_w(x) \sim x^{-w} \sum_{k=0}^{\infty} [-x]^{-kw} \prod_{l=1}^{k} \frac{\Gamma(1 - lw)}{\Gamma(\frac{1}{2} - lw)}$$

is more useful for large x. The so-called Koutecky function of polarography (Delahay, 1954) is simply

$$F(y) = \tfrac{1}{2} y f_{3/14}([y/2]^{14/3}).$$

This function describes the current passed by an expanding spherical electrode under control by both diffusion and kinetics.

CHAPTER 9
REPRESENTATION OF TRANSCENDENTAL FUNCTIONS

The message of this ninth chapter is that the fractional calculus enables a plethora of transcendental functions to be expressed in terms of a basis set of only three functions: the inverse binomial, the exponential, and the zero-order Bessel. Using only simple algebraic operations (multiplication by constants and powers) in addition to differintegration, each of these reducible transcendentals may be synthesized from one or other of the three basis functions. A partial listing of reducible transcendentals includes: exponential-related functions such as exponential integrals and error functions; logarithms; inverse trigonometric functions and their hyperbolic counterparts; incomplete gamma and beta functions; circular and hyperbolic sines, cosines, Fresnel integrals, sine integrals, cosine integrals, Bessel functions, and Struve functions; generalized functions such as Kummer, Gauss and other hypergeometric functions; Legendre functions and associated Legendre functions; and elliptic integrals.

Our position is that these differintegral representations of transcendental functions constitute more appealing definitions than those customary in terms of definite integrals, differential equations, and the like. Indeed, we are tempted to go even further and assert that differintegration removes the need to recognize many transcendentals as functions in their own right. Thus, if the incomplete gamma function is nothing but the differintegral of an exponential (as Section 6.2 shows it to be), is any purpose served in regarding it as anything but the differintegral of an exponential? Need it be graced with a special name? Our viewpoint is that, in this case (as in many others), the description "exponential differintegral" and the symbol $\exp(-x) d^q\{\exp(x)\}/dx^q$ are more informative than the name "incomplete gamma function" and the symbol $\gamma^*(-q, x)$.

The definition of and our notation for a generalized hypergeometric function were presented in Section 2.10; the behavior of these functions under

9 REPRESENTATION OF TRANSCENDENTAL FUNCTIONS

differintegration was derived in Section 6.6; and the reader will have met these useful generalized functions elsewhere in this text. In this chapter we shall encounter many instances of the product of a power of x and a generalized hypergeometric function; that is,

$$x^p \left[\pm x \frac{b_1, b_2, \ldots, b_K}{c_1, c_2, \ldots, c_L} \right], \qquad p > -1.$$

For brevity we shall term such products simply "hypergeometrics." They constitute examples of differintegrable series (see Section 3.1).

9.1 TRANSCENDENTAL FUNCTIONS AS HYPERGEOMETRICS

The majority of those important transcendental functions which remain finite[1] as their arguments approach zero are expressible in terms of generalized hypergeometric functions of complexity $\frac{0}{0}$, $\frac{0}{1}$, $\frac{1}{1}$, $\frac{0}{2}$, $\frac{1}{2}$, $\frac{2}{2}$, or $\frac{1}{3}$. This section is devoted merely to illustrating this statement by citing copious examples.

We shall find it convenient to denote by $\pm x$ the argument of the hypergeometric function; accordingly the argument of the transcendental function is not always x (frequently being $\sqrt{x}$ or $2\sqrt{x}$). A change of independent variable is, of course, always possible. Usually, the formulas below imply $x > 0$; sometimes the restriction $0 < x < 1$ is required. Where the left-hand member of the formula incorporates two functions, such as $\frac{\sinh(x)}{\sin(x)}$, the upper formula is associated with a positive argument $+x$ of the hypergeometric function, the lower with the negative argument $-x$.

The only instances of hypergeometric functions of $\frac{0}{0}$ complexity are

(9.1.1) $$\frac{1}{1 \mp x} = \left[\pm x \, \frac{}{} \right].$$

Hypergeometrics of $\frac{0}{1}$ complexity are exponential-like. The general case is

$$\exp(x) \gamma^*(c, x) = \left[x \, \frac{}{c} \right],$$

---

[1] Functions unbounded at $x = 0$ may also be expressed as hypergeometrics provided the singularity at zero is a branch point of order less than unity.

9.1 TRANSCENDENTAL FUNCTIONS AS HYPERGOEMETRICS

but the important instances,

$$x^n \exp(x) = \left[x \frac{}{-n} \right], \qquad n = 0, 1, 2, \ldots,$$

$$\exp(x) \operatorname{erf}(\sqrt{x}) = \sqrt{x} \left[x \frac{}{\tfrac{1}{2}} \right],$$

and Dawson's integral

$$\operatorname{daw}(\sqrt{x}) = \frac{\sqrt{\pi}}{2} \left[-x \frac{}{\tfrac{1}{2}} \right],$$

are included therein.

Instances of the general case

$$\frac{B_x(c, b - c + 1)}{[1 - x]^{b-c+1}} = \frac{\Gamma(c) x^c}{\Gamma(b+1)} \left[x \frac{b}{c} \right]$$

of a $\tfrac{1}{1}$ hypergeometric are numerous. They include the logarithm[2] (Section 6.7) as well as binomials

$$\frac{1}{[1 \mp x]^{b+1}} = \frac{1}{\Gamma(b+1)} \left[\pm x \frac{b}{0} \right]$$

and such inverse trigonometric functions as

$$\frac{\arcsin(\sqrt{x})}{\sqrt{1-x}} = \frac{\sqrt{\pi x}}{2} \left[x \frac{0}{\tfrac{1}{2}} \right]$$

and hyperbolic functions as

$$\frac{\operatorname{arctanh}(\sqrt{x})}{\arctan(\sqrt{x})} = \left[\pm x \frac{-\tfrac{1}{2}}{\tfrac{1}{2}} \right].$$

Oscillatory functions of argument $2\sqrt{x}$ and their hyperbolic counterparts are mostly hypergeometrics of $\tfrac{0}{2}$ complexity. Thus we find the sines,

$$\frac{\sinh(2\sqrt{x})}{\sin(2\sqrt{x})} = \sqrt{\pi x} \left[\pm x \frac{}{0, \tfrac{1}{2}} \right],$$

the cosines,

$$\frac{\cosh(2\sqrt{x})}{\cos(2\sqrt{x})} = \sqrt{\pi} \left[\pm x \frac{}{-\tfrac{1}{2}, 0} \right],$$

---

[2] The generalized logarithm that will be discussed in Section 10.5 is also a hypergeometric of $\tfrac{1}{1}$ complexity.

the Bessel functions,

$$\frac{I_\nu(2\sqrt{x})}{J_\nu(2\sqrt{x})} = x^{\frac{1}{2}\nu}\left[\pm x \frac{}{0, \nu}\right],$$

and the Struve functions,

(9.1.2) $$\frac{L_\nu(2\sqrt{x})}{H_\nu(2\sqrt{x})} = x^{\frac{1}{2}+\frac{1}{2}\nu}\left[\pm x \frac{}{\frac{1}{2}, \frac{1}{2}+\nu}\right],$$

to be examples of this general pattern.

Closely related are the Airy functions, of which

$$\text{fai}([9x]^{\frac{1}{3}}) = \Gamma(\tfrac{2}{3})\left[x \frac{}{-\tfrac{1}{3}, 0}\right]$$

and

$$\text{gai}([9x]^{\frac{1}{3}}) = \Gamma(\tfrac{1}{3})[\tfrac{1}{3}x]^{\frac{1}{3}}\left[x \frac{}{0, \tfrac{1}{3}}\right]$$

are examples. We here use fai( ) and gai( ) to denote those Airy functions that Abramowitz and Stegun (1964, Section 10.4) denote by $f(\ )$ and $g(\ )$.

A variety of transcendental functions comprise the hypergeometric class of $\frac{1}{2}$ complexity. Those with one denominatorial parameter equal to zero are Kummer functions (Section 2.10), of which

$$\text{erf}(\sqrt{x}) = \sqrt{\frac{x}{\pi}}\left[-x \frac{-\tfrac{1}{2}}{0, \tfrac{1}{2}}\right]$$

provides one example, many others being listed by Abramowitz and Stegun (1964, p. 509). An illustration of a $\frac{1}{2}$ hypergeometric that is not a Kummer function is the function

$$\text{Ei}(x) - \ln(x) = \gamma + x\left[x \frac{0}{1, 1}\right]$$

related to the exponential integral.

An example of a hypergeometric of $\frac{2}{2}$ complexity is the associated Legendre function

(9.1.3) $$\frac{\Gamma(\tfrac{1}{2}\nu - \tfrac{1}{2}\mu + \tfrac{1}{2})\Gamma(-\tfrac{1}{2}\nu - \tfrac{1}{2}\mu)}{2^\mu} P_\nu^\mu(\sqrt{1-x})$$
$$= x^{-\tfrac{1}{2}\mu}\left[x \frac{\tfrac{1}{2}\nu - \tfrac{1}{2}\mu - \tfrac{1}{2}, -\tfrac{1}{2}\nu - \tfrac{1}{2}\mu - 1}{-\mu, 0}\right].$$

This hypergeometric function has two disposable constants v and μ in addition to the independent variable x. Other important $\frac{2}{2}$ hypergeometrics, such as the elliptic integral

$$K(x) = \frac{1}{2}\left[x \; \frac{-\frac{1}{2}, -\frac{1}{2}}{0, 0}\right]$$

and the inverse hyperbolic sine

(9.1.4) $\qquad \operatorname{arcsinh}(\sqrt{x}) = \frac{1}{2}\sqrt{\frac{x}{\pi}}\left[-x \; \frac{-\frac{1}{2}, -\frac{1}{2}}{0, \frac{1}{2}}\right],$

have no such disposable constants. If at least one of the denominatorial parameters is zero, as in the three examples cited above, the hypergeometric is an example of the Gauss function [see Section 2.10 and Abramowitz and Stegun (1964, Chap. 15)] about which a large body of theory exists.

The greatest complexity of hypergeometrics that we here consider is $\frac{1}{3}$. The sine integral and its hyperbolic analog provide examples,

(9.1.5) $\qquad \begin{aligned}\operatorname{Shi}(2\sqrt{x}) \\ \operatorname{Si}(2\sqrt{x})\end{aligned} = \left[\pm x \; \frac{-\frac{1}{2}}{0, \frac{1}{2}, \frac{1}{2}}\right].$

9.2 HYPERGEOMETRICS WITH $K > L$

We have chosen not to stress the convergence aspects of hypergeometric functions. From the $j \to \infty$ properties of x^j and $\Gamma(j + \text{constant})$, however, we would expect the following rule to apply:

$K < L$: convergence for all finite x,
$K = L$: convergence for $-1 \leq x < 1$ only,
$K > L$: divergence for all x except $x = 0$.

And, indeed, this rule does seem generally applicable.

Because it represents a series which diverges for all nontrivial values of its argument, one might imagine that any hypergeometric function for which the number K of numeratorial parameters exceeds the number L of denominatorial parameters would be a useless concept. This is not quite the case because such a hypergeometric may asymptotically represent an important transcendental function. For example, we have the asymptotic relation

$$\exp\left(\frac{1}{x}\right) E_1\left(\frac{1}{x}\right) \sim x\left[-x \; \frac{0}{\quad}\right],$$

valid as $x \to 0$. Here $E_1(\ )$ is the exponential integral, defined by

$$E_1(x) = \int_x^\infty \frac{\exp(-y)\,dy}{y}.$$

Some similar relations are

$$\exp\left(\frac{1}{x}\right)\mathrm{erfc}\left(\frac{1}{\sqrt{x}}\right) \sim \frac{\sqrt{x}}{\pi}\left[-x \frac{-\tfrac{1}{2}}{}\right], \qquad x \to 0,$$

and

$$\tfrac{1}{2}\pi \sin(x) - \mathrm{Si}(x)\sin(x) - \mathrm{Ci}(x)\cos(x) \sim \frac{2x^2}{\sqrt{\pi}}\left[-4x^{-2}\frac{0,\tfrac{1}{2}}{}\right],$$

but we shall find no further use for such asymptotic representations.

9.3 REDUCTION OF COMPLEX HYPERGEOMETRICS

In Section 6.6 it was demonstrated that the formula

$$\frac{d^q}{dx^q}\left\{x^p\left[\beta x \frac{b_1, b_2, \ldots, b_K}{c_1, c_2, \ldots, c_L}\right]\right\} = x^{p-q}\left[\beta x \frac{p, b_1, b_2, \ldots, b_K}{p-q, c_1, c_2, \ldots, c_L}\right]$$

expresses the result of differintegrating a power hypergeometric product of complexity $\frac{K}{L}$, provided p exceeds -1. Because of the cancellation property of hypergeometric parameters (Section 2.10), it follows that the sequence of operations

(i) multiplication by x^{c-p},
(ii) differintegration to order $c - b$,
(iii) multiplication by x^{p-b},

where b and c are respectively a numeratorial and a denominatorial parameter, accomplishes the transformation

$$x^{P-b}\frac{d^{c-b}}{dx^{c-b}}\left\{x^{c-P}x^P\left[\beta x \frac{b, b_2, \ldots, b_K}{c, c_2, \ldots, c_L}\right]\right\} = x^P\left[\beta x \frac{b_2, \ldots, b_K}{c_2, \ldots, c_L}\right],$$

converting one hypergeometric of complexity $\frac{K}{L}$ to another of complexity $\frac{K-1}{L-1}$. For example, setting $\beta = 1$, $K = 1$, $b = -\tfrac{1}{2}$, $L = 3$, $c = 0$, $c_2 = c_3 = \tfrac{1}{2}$, and $p = P = \tfrac{1}{2}$ gives

(9.3.1) $$x\frac{d^{\tfrac{1}{2}}}{dx^{\tfrac{1}{2}}}\left\{x^{-\tfrac{1}{2}}\sqrt{x}\left[x\frac{-\tfrac{1}{2}}{0,\tfrac{1}{2},\tfrac{1}{2}}\right]\right\} = \sqrt{x}\left[x\frac{}{\tfrac{1}{2},\tfrac{1}{2}}\right].$$

9.3 REDUCTION OF COMPLEX HYPERGEOMETRICS

The procedure may be iterated to enable a $\frac{K}{L}$ hypergeometric to be reduced in complexity to $\frac{K-n}{L-n}$. An example is

(9.3.2) $\quad \dfrac{d}{dx}\left\{x\dfrac{d^{\frac{1}{2}}}{dx^{\frac{1}{2}}}\left\{x^{-\frac{1}{2}}\sqrt{x}\left[-x\;\dfrac{-\frac{1}{2},\,-\frac{1}{2}}{0,\,\frac{1}{2}}\right]\right\}\right\} = \left[-x\;\underline{\quad\quad}\right].$

The value of such expressions is that they allow an interrelation between important transcendental functions. Thus, the hypergeometrics in equation (9.3.1) will be recognized from Section 9.1 as the hyperbolic sine integral and the zero-order Struve function [see (9.1.5) and (9.1.2), respectively], so that

$$x\frac{d^{\frac{1}{2}}}{dx^{\frac{1}{2}}}\{x^{-\frac{1}{2}}\operatorname{Shi}(2\sqrt{x})\} = L_0(2\sqrt{x}).$$

Likewise,

$$2\sqrt{\pi}\,\frac{d}{dx}\left\{x\frac{d^{\frac{1}{2}}}{dx^{\frac{1}{2}}}\{x^{-\frac{1}{2}}\operatorname{arcsinh}(\sqrt{x})\}\right\} = \frac{1}{1+x}$$

follows upon recognition of the hypergeometrics on each side of equation (9.3.2) from expressions (9.1.4) and (9.1.1).

It is evident that K applications of the sequence (i) through (iii) will convert a $\frac{K}{L}$ hypergeometric to one of $\frac{0}{L-K}$ complexity. Such a hypergeometric has the general form

$$x^p\left[\beta x\;\frac{}{c_1,\,c_2,\,\ldots,\,c_{L-K}}\right],$$

being without numeratorial parameters. The sequence of operations

(a) multiplication by x^{c-p},
(b) differintegration to order c,

where c is any denominatorial parameter, now serves to replace that parameter by a zero, thus:

$$\frac{d^c}{dx^c}\left\{x^{c-p}x^p\left[\beta x\;\frac{}{c,\,c_2,\,\ldots,\,c_{L-K}}\right]\right\} = \left[\beta x\;\frac{}{0,\,c_2,\,\ldots,\,c_{L-K}}\right].$$

Hence, if this sequence is repeated $L-K$ times, all the denominatorial parameters may be replaced by zeros.

In summary then: By a sufficient number of operations, each of which is either multiplication by a power of x or differintegration with respect to x, any hypergeometric of complexity $\frac{K}{L}$ is reducible to one of complexity $\frac{0}{L-K}$ in which all denominatorial parameters are zero.

9.4 BASIS HYPERGEOMETRICS

As we have just seen, those generalized hypergeometric functions that are without numeratorial parameters and whose denominatorial parameters are all zero play a distinguished role in that they are the end result of the reduction of all other hypergeometrics. We shall refer to them as "basis hypergeometrics." The three most important are

$$\left[x \longrightarrow \right] = \frac{1}{1-x}, \qquad -1 < x < 1,$$

$$\left[x \xrightarrow{} 0 \right] = \exp(x), \qquad -\infty < x < \infty,$$

and

$$\left[x \xrightarrow{} 0,0 \right] = I_0(2\sqrt{x}), \qquad 0 \leq x < \infty.$$

Figure 9.4.1 compares these three basis functions on the interval $0 \leq x < 1$, the only range which they share in common. Also shown on the figure are the complementary functions

$$\left[-x \longrightarrow \right] = \frac{1}{1+x},$$

$$\left[-x \xrightarrow{} 0 \right] = \exp(-x),$$

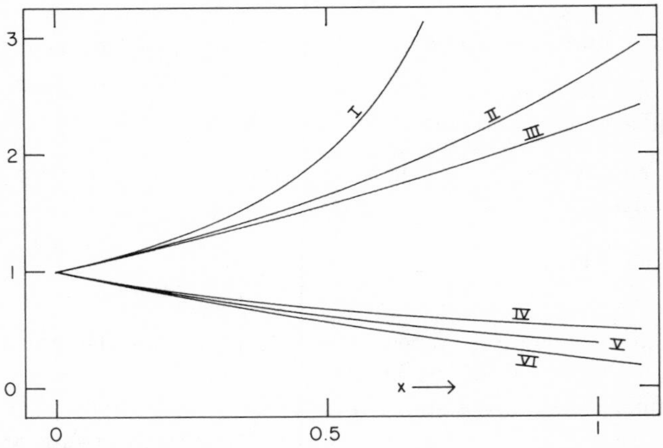

FIG. 9.4.1. The three most important basis functions: I = $[1-x]^{-1}$, II = $\exp(x)$, and III = $I_0(2\sqrt{x})$; and their complements: IV = $[1+x]^{-1}$, V = $\exp(-x)$, and VI = $J_0(2\sqrt{x})$.

and

$$\left[-x \frac{}{0,0}\right] = I_0(2i\sqrt{x}) = J_0(2\sqrt{x}).$$

Reversing the process discussed in the preceding section, it is possible by employing only the operations of multiplication by x^p and differintegration to order q to generate from each basis hypergeometric a host[3] of transcendental functions. The next three sections will be devoted to detailing such function syntheses. In those sections we shall make use of a special kind of chart, a "function synthesis diagram," and it is appropriate to set the stage for these future sections by explaining here the significance of these diagrams.

In a function synthesis diagram, each hypergeometric occupies a single[4] point in a plane. The Cartesian coordinates of the point are $(p + 2B - 2C, p)$ where p is the exponent of the power, B is the sum $[= \sum b_k]$ of the numeratorial parameters of the generalized hypergeometric function, and C is is the sum $[= \sum c_l]$ of the denominatorial parameters. The origin $(0, 0)$ is occupied by the basis hypergeometric, but, in common with all points in the plane, the origin is also the location of an infinity of other functions. Thus, for example, the hypergeometrics

$$x^{\frac{1}{2}}\left[x \frac{\frac{1}{2}}{1}\right], \quad x^{\frac{1}{2}}\left[x \frac{-\frac{1}{2}, \frac{1}{2}}{-1, \frac{3}{2}}\right], \quad \text{and} \quad x^{\frac{1}{2}}\left[x \frac{-\frac{1}{2}}{0}\right]$$

are all colocated at $(-\frac{1}{2}, \frac{1}{2})$. To depict function syntheses in such a diagram, points representing hypergeometrics are connected by diagonal arrows. An arrow thus ↗ (always at a $\pi/4$ angle) indicates multiplication by a positive power of x. Similarly, division by a positive power of x is indicated by ↙. Differintegration is represented by arrows at right angles to these: upwards ↖ in the case of integration (to fractional or integer order) and downwards ↘ for differentiation. The index (power p in x^p or order q in d^q/dx^q) may be indicated in a "box" on the arrow. The example, Fig. 9.4.2, illustrates most of the features of a function synthesis diagram.

---

[3] But not quite every hypergeometric! The recurrences displayed in Section 2.10 show that two hypergeometrics of the same complexity $\frac{K}{L}$ differ by a term that may vanish on differintegration. This possible coalescence of two hypergeometrics on differintegration implies that not every hypergeometric may be regenerated from the appropriate basis function.

[4] Cyclodifferential functions (Section 6.11) will occupy more than one point.

9 REPRESENTATION OF TRANSCENDENTAL FUNCTIONS

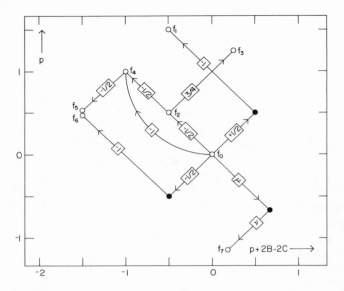

FIG. 9.4.2. Synthesis diagram for seven fictitious functions.

Figure 9.4.2 shows how seven fictitious functions may be generated from a basis hypergeometric f_0. Function f_1 is synthesized by multiplying f_0 by $x^{+1/2}$ and integrating the product, i.e.,

$$f_1 = \frac{d^{-1}}{dx^{-1}}\{\sqrt{x}f_0\}.$$

Function f_2 is the semiintegral of f_0, while

$$f_3 = x^{3/4}f_2 = x^{3/4}\frac{d^{-1/2}f_0}{dx^{-1/2}}$$

is the product of $x^{3/4}$ and the semiintegral of f_0. Notice that though the synthetic routes to f_1 and f_3 both traverse the point $(0, 1)$, they will, in general, not be coincident there because the operations of multiplication and differintegration do not commute. Function f_4 is the semiintegral of f_2 and on account of the composition law (Section 5.7) it is also the integral of f_0. Hence two pathways —two semiintegrations, or a single integration (shown by the looped arrow) —produce f_4 from f_0; indeed, four quarter integrations would also serve. Normally only one of these alternatives would be included in a function synthesis diagram. Because of the lack of commutativity, two routes around a rectangle do not normally produce the same function. Thus, even though they occupy the same site $(-\frac{3}{2}, \frac{1}{2})$ in the plane, functions f_5 (equal to $f_4/\sqrt{x}$)

9.4 BASIS HYPERGEOMETRICS

and f_6 (equal to $d^{-1}\{f_0/\sqrt{x}\}/dx^{-1}$) are distinct. Function f_7 is synthesized from the basis hypergeometric by differintegration to order μ, followed by multiplication by x^ν; that is,

$$f_7 = x^\nu \frac{d^\mu f_0}{dx^\mu},$$

where, as drawn, μ is positive and ν negative. Notice that f_7 lies below the $p = -1$ line; since it is not generally possible to differintegrate such hypergeometrics, f_7 is not a suitable starting point for further synthetic steps.

The numbers shown in boxes on Fig. 9.4.2 will be seen often to be redundant, the same information being conveyed by the lengths of the arrows and also by their projection on either axis. Accordingly, we shall often omit them from function synthesis diagrams whenever their omission leaves no danger of confusion.

Before leaving the subject of basis hypergeometrics, we wish to call attention to an interesting interrelation which exists between the inverse binomial function, the exponential function, and the zero-order Bessel function of argument $2\sqrt{x}$, that is, between the most important basis hypergeometrics. It may already have been noted by the reader who is an expert in Laplace transformation that

$$\mathscr{L}\{I_0(2\sqrt{x})\} = \frac{1}{s}\exp\left(\frac{1}{s}\right)$$

and

$$\mathscr{L}\{\exp(x)\} = \frac{1}{s-1} = \frac{1}{s}\left[\frac{1}{1-[1/s]}\right].$$

Thus Laplace transformation converts a basis hypergeometric into a hypergeometric of lower complexity. These transforms are just examples of the general rule

$$\mathscr{L}\left\{\left[x\,\frac{b_1, b_2, \ldots, b_K}{c_1, c_2, \ldots, c_L}\right]\right\} = \frac{1}{s}\left[\frac{1}{s}\,\frac{0, b_1, b_2, \ldots, b_K}{c_1, c_2, \ldots, c_L}\right]$$

for Laplace transformation of a generalized hypergeometric function. By invoking this relationship it is possible to Laplace transform any function that can be expressed as a hypergeometric,

$$\mathscr{L}\left\{x^p\left[\beta x\,\frac{b_1, \ldots, b_K}{c_1, \ldots, c_L}\right]\right\} = \frac{1}{s^{p+1}}\left[\frac{\beta}{s}\,\frac{0, 0, b_1, \ldots, b_K}{p, c_1, \ldots, c_L}\right].$$

172 9 REPRESENTATION OF TRANSCENDENTAL FUNCTIONS

More importantly, it is also possible to Laplace invert any transform which is expressible as a hypergeometric of s^{-1}, thus:

$$\mathscr{L}^{-1}\left\{\frac{1}{s^p}\left[\frac{\beta}{s}\,\frac{b_1,\ldots,b_K}{c_1,\ldots,c_L}\right]\right\} = x^{p-1}\left[\beta x\,\frac{p-1, b_1,\ldots,b_K}{0,0,c_1,\ldots,c_L}\right].$$

9.5 SYNTHESIS OF $K = L$ TRANSCENDENTALS

Transcendentals that are representable by a hypergeometric having equal numbers of numeratorial and denominatorial parameters may be synthesized from the basis hypergeometric $[1 - x]^{-1}$ (or its $[1 + x]^{-1}$ complement). We shall give some examples in the form of function synthesis diagrams, omitting proofs. Proofs can always be established by replacing the functions by their hypergeometric equivalents (many of which will be found listed in Section 9.1) and invoking equation (6.6.3).

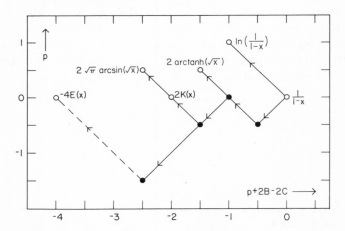

FIG. 9.5.1. Four transcendental functions synthesized from the $[1 - x]^{-1}$ basis. $E(x)$ cannot be synthesized via the dashed route.

Figure 9.5.1 is a function synthesis diagram showing how four transcendentals may be generated from the basis hypergeometric $[1 - x]^{-1}$. Similar syntheses starting with $[1 + x]^{-1}$ would have produced $\ln(1 + x)$, $2\arctan(\sqrt{x})$, $2K(-x)$,[5] and $2\sqrt{\pi}\,\text{arcsinh}(\sqrt{x})$. Since

$$(9.5.1) \qquad -4E(x) = \left[x\,\frac{-\tfrac{3}{2}, -\tfrac{1}{2}}{0, 0}\right],$$

[5] An alternative representation of the complete elliptic integral $K(-x)$ is $[1 + x]^{-\tfrac{1}{2}} K(x/[1 + x])$.

this elliptic integral can, formally, be generated by the synthetic route indicated by the dotted arrow. However, this route involves differintegration of a term proportional to $x^{-3/2}$, an invalid operation. To generate $E(x)$ legitimately, we must make use of the identity

$$2\pi - 4E(x) = x\left[x \frac{-\tfrac{1}{2},\tfrac{1}{2}}{1,1}\right],$$

which follows from (9.5.1) and one of the recursions cited in Section 2.10. One of the two[6] synthetic routes to this hypergeometric is mapped out in Fig. 9.5.2. The route is four-legged, consisting serially of multiplication by $\sqrt{x}$, semiintegration, division by x^{3}, and finally sesquiintegration.

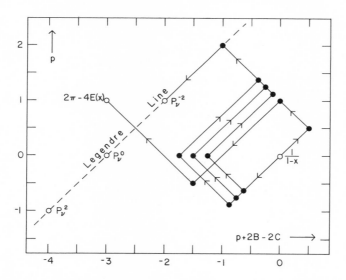

FIG. 9.5.2. All associated Legendre functions lie on the dashed line and can be synthesized from the $[1-x]^{-1}$ basis by a five-legged route akin to the ones shown.

Figure 9.5.2. will also be used to illustrate some of the properties of the associated Legendre function $P_\nu^\mu(\sqrt{1-x})$. Note first of all, from the equation (9.1.3) which gives the hypergeometric representation of this function, that its parameter difference $B - C$ equals $-\tfrac{1}{2}$ and is independent of both μ and ν. In terms of our function synthesis diagram, this means that all associated Legendre functions lie on the line annotated "Legendre line" in Fig. 9.5.2. Moreover, since p for these functions equals $-\mu/2$ and is ν-independent, it

---

[6] There will, in general, be four nonequivalent routes from a $\tfrac{0}{0}$ to a $\tfrac{2}{2}$ hypergeometric, but a redundancy enters into the synthesis of $E(x)$ because of the identity of the two denominatorial parameters.

follows that μ alone determines the position on the Legendre line at which a given associated Legendre function is located. For example all associated Legendre functions of orders 2, 0, and -2 are located at the points marked $P_\nu{}^2$, $P_\nu{}^0$, and $P_\nu{}^{-2}$ on the chart.[7] In synthesizing these functions from the $[1-x]^{-1}$ basis hypergeometric, it is the route alone which determines ν, the starting and ending points being independent of the degree ν. The chart shows three five-legged routes from $[1-x]^{-1}$ to the point labelled $P_\nu{}^{-2}$, at which the functions $4\Gamma(\tfrac{3}{2}+\tfrac{1}{2}\nu)\Gamma(1-\tfrac{1}{2}\nu)P_\nu{}^{-2}(\sqrt{1-x})$ are located. The routes form a family that differs only in the ν value of the associated Legendre function which arises from the five-legged route. From the infinity of possible ν values, three only ($\nu = \tfrac{5}{4}, \tfrac{3}{2}$, and $\tfrac{7}{4}$) are shown on Fig. 9.5.2.

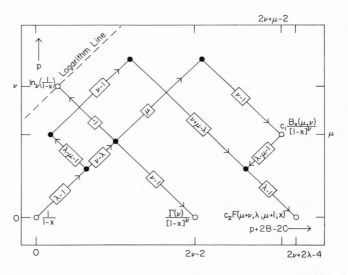

FIG. 9.5.3. Synthesis diagram for some very general $K = L$ functions. All generalized logarithms lie on the dashed line. The constants c_1 and c_2 that multiply the synthesized incomplete beta and Gauss functions are, respectively, $\Gamma(\mu+\nu)/\Gamma(\nu)$ and $\Gamma(\mu+1)/[\Gamma(\lambda)\Gamma(\mu+\nu)]$.

Three very general functions—the inverse binomial, the incomplete beta function, and the Gauss hypergeometric function—are synthesized via routes shown in Fig. 9.5.3. These functions have, respectively, one, two, and three

---

[7] The associated Legendre functions of zero order are usually notated P_ν rather than $P_\nu{}^0$ and are termed Legendre functions (or Legendre polynomials if the degree ν is an integer).

disposable parameters. The route to a generalized logarithm (Section 10.5) is also mapped out in this chart. These latter functions all lie on the line marked "logarithm line."

9.6 SYNTHESIS OF $K = L - 1$ TRANSCENDENTALS

Figure 9.6.1 is a synthesis diagram showing how four transcendental functions are derived from the basis $K = L - 1$ hypergeometric, $\exp(x)$. Starting with the complementary basis function $\exp(-x)$, similar routes lead to $\sqrt{\pi}\exp(-\frac{1}{2}x)I_1(\frac{1}{2}x)$, $E_1(x) + \ln(x) + \gamma$, and $\sqrt{\pi}\exp(-\frac{1}{2}x)I_0(\frac{1}{2}x)$.

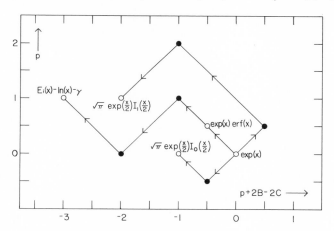

FIG. 9.6.1. Some functions synthesized from the $\frac{0}{1}$ basis hypergeometric function $\exp(x)$.

However, some additional important functions—Dawson's integral, the error function, and the error function complement integral—also arise from $\exp(-x)$, as shown in Fig. 9.6.2.

There are two important multivariate functions which are $K = L - 1$ hypergeometrics. The first is the incomplete gamma function,

$$\gamma^*(c, x) = \frac{1}{\Gamma(c)}\left[-x \frac{c-1}{0, c}\right] = \frac{x^{-c}}{\Gamma(c)}\frac{d^{-1}}{dx^{-1}}\{x^{c-1}\exp(-x)\},$$

the second being the Kummer confluent function $M(\ ,\ ,\ )$. Figure 9.6.3 shows synthetic routes to these functions from $\exp(x)$.

176   9 REPRESENTATION OF TRANSCENDENTAL FUNCTIONS

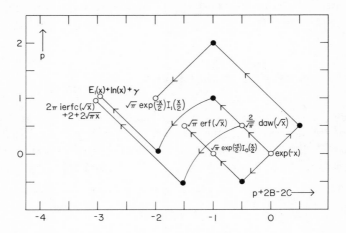

FIG. 9.6.2. Some functions synthesized from the complementary $\frac{0}{1}$ basis function $\exp(-x)$.

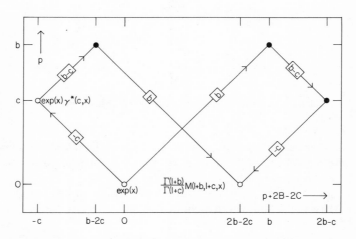

FIG. 9.6.3. Additional functions that may be synthesized from $\exp(x)$.

9.7 SYNTHESIS OF $K=L-2$ TRANSCENDENTALS

Most of the important transcendental functions of mathematical physics belong to the $K = L - 2$ class of hypergeometrics and may be synthesized from the basis function $I_0(2\sqrt{x})$ or its complementary $J_0(2\sqrt{x})$.

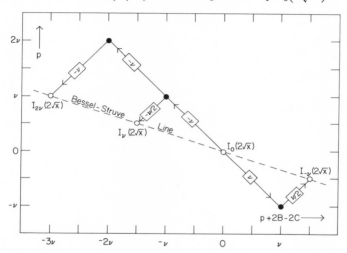

FIG. 9.7.1. Synthesis of Bessel functions. All Bessel and Struve functions of argument $2\sqrt{x}$ lie on the dashed line.

First consider the synthesis of a Bessel function $I_\nu(2\sqrt{x})$ of nonzero order from the basis hypergeometric. As depicted in Fig. 9.7.1, this synthesis is accomplished by what a chess player might term a "knight's move," the first leg of the route being twice as long as the second. From this it follows that all Bessel functions are located in our function synthesis plane on a line through the origin and inclined at an angle $\arctan(-\frac{1}{3})$. Next, consider the hypergeometric representation

$$I_{\mu+\frac{1}{2}}(2\sqrt{x}) = x^{\frac{1}{2}\mu+\frac{1}{4}} \left[x \frac{}{0, \mu + \frac{1}{2}} \right]$$

of a Bessel function of order $\mu + \frac{1}{2}$, compared with that of a Struve function

$$L_{\mu-\frac{1}{2}}(2\sqrt{x}) = x^{\frac{1}{2}\mu+\frac{1}{4}} \left[x \frac{}{\frac{1}{2}, \mu} \right]$$

of order $\mu - \frac{1}{2}$. Their p and $B - C$ values being identical, it is evident that $I_{\mu+\frac{1}{2}}(2\sqrt{x})$ and $L_{\mu-\frac{1}{2}}(2\sqrt{x})$ must be colocated on what we shall henceforth

9 REPRESENTATION OF TRANSCENDENTAL FUNCTIONS

term the Bessel–Struve line. It is equally evident, however, that $I_{\mu+\frac{1}{2}}(2\sqrt{x})$ and $L_{\mu-\frac{1}{2}}(2\sqrt{x})$ are distinct functions for all μ except $\mu = 0$. For this unique value, the Bessel and Struve functions coalesce:

$$I_{\frac{1}{2}}(2\sqrt{x}) = L_{-\frac{1}{2}}(2\sqrt{x}) = \frac{\sinh(2\sqrt{x})}{\sqrt{\pi}\, x^{\frac{1}{4}}}.$$

The Bessel–Struve line also appears on Fig. 9.7.2, with the synthetic routes to as many as nine important transcendentals, starting from the basis hypergeometric $I_0(2\sqrt{x})$. Of course, similar synthetic routes from the complementary $J_0(2\sqrt{x})$ basis will yield the corresponding circular functions, $\sin(2\sqrt{x})/\sqrt{\pi}$, $\cos(2\sqrt{x})/\sqrt{\pi}$, and $\mathrm{Si}(2\sqrt{x})$, as well as J Bessel functions and H Struve functions.

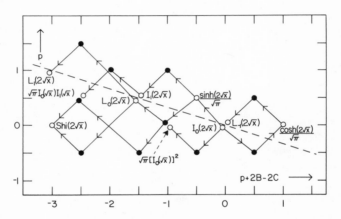

FIG. 9.7.2. Synthetic routes from the $\frac{0}{2}$ basis hypergeometric function $I_0(2\sqrt{x})$.

Notice the interesting synthetic relationship between the basis hypergeometric $I_0(2\sqrt{x})$ and the square of the closely related $I_0(\sqrt{x})$ function. This may be derived starting from the well-known binomial coefficient summation

$$\sum_{k=0}^{n} \binom{n}{k}^2 = \binom{2n}{n}$$

[a special case of equation (1.3.20)] that may be rewritten in gamma function notation as

$$\sum_{k=0}^{n} \frac{1}{[\Gamma(k+1)]^2 [\Gamma(n-k+1)]^2} = \frac{\Gamma(2n+1)}{[\Gamma(n+1)]^4}.$$

9.7 SYNTHESIS OF K = L − 2 TRANSCENDENTALS   179

If, now, the duplication formula (1.3.9) is employed to expand the $\Gamma(2n+1)$ term, and each side of the equation is multiplied by $[\tfrac{1}{4}x]^n$,

$$\sum_{k=0}^{n} \frac{[\tfrac{1}{4}x]^k}{[\Gamma(k+1)]^2} \frac{[\tfrac{1}{4}x]^{n-k}}{[\Gamma(n-k+1)]^2} = \frac{x^n \Gamma(n+\tfrac{1}{2})}{\sqrt{\pi}\,[\Gamma(n+1)]^3}$$

results. The natural number n may now be regarded as an index, and the summation from $n = 0$ to $n = \infty$ carried out. With use of the

$$\sum_{n=0}^{\infty} \sum_{k=0}^{n} \equiv \sum_{k=0}^{\infty} \sum_{n-k=0}^{\infty}$$

permutation, this becomes

$$\sum_{k=0}^{\infty} \frac{[\tfrac{1}{4}x]^k}{[\Gamma(k+1)]^2} \sum_{n-k=0}^{\infty} \frac{[\tfrac{1}{4}x]^{n-k}}{[\Gamma(n-k+1)]^2} = \frac{1}{\sqrt{\pi}} \sum_{n=0}^{\infty} \frac{x^n \Gamma(n+\tfrac{1}{2})}{[\Gamma(n+1)]^3},$$

which is easily expressed as

$$\left[\tfrac{1}{4}x \frac{}{0,0}\right]\left[\tfrac{1}{4}x \frac{}{0,0}\right] = \frac{1}{\sqrt{\pi}}\left[x \frac{-\tfrac{1}{2}}{0,0,0}\right]$$

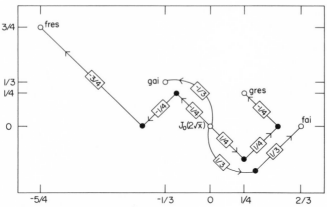

FIG. 9.7.3. A synthesis diagram involving small steps. The abbreviations

$$\text{gai} = 3^{\tfrac{1}{3}}\,\text{gai}(-[9x]^{\tfrac{1}{3}})/\Gamma(\tfrac{1}{3}),$$
$$\text{fai} = \text{fai}(-[9x]^{\tfrac{1}{3}})/\Gamma(\tfrac{2}{3}),$$
$$\text{fres} = \frac{1}{2\sqrt{\pi}}\,\text{fres}\!\left(\frac{2x^{\tfrac{1}{4}}}{\sqrt{\pi}}\right) - \frac{1}{\sqrt{\pi}}\cos(2\sqrt{x}+\tfrac{1}{4}\pi),$$

and

$$\text{gres} = \frac{1}{\sqrt{\pi}}\sin(2\sqrt{x}+\tfrac{1}{4}\pi) - \sqrt{\frac{2}{\pi}}\,\text{gres}\!\left(\frac{2x^{\tfrac{1}{4}}}{\sqrt{\pi}}\right)$$

are employed.

9 REPRESENTATION OF TRANSCENDENTAL FUNCTIONS

in terms of hypergeometric functions. From here it is only a short step to

(9.7.1) $$\frac{d^{-\frac{1}{2}}}{dx^{-\frac{1}{2}}}\left\{\frac{I_0(2\sqrt{x})}{\sqrt{x}}\right\} = [I_0(\sqrt{x})]^2,$$

as portrayed in Fig. 9.7.2. Reminiscent of the classical result

$$\frac{d^{-1}}{dx^{-1}}\{\sin(2x)\} = [\sin(x)]^2,$$

equation (9.7.1) is one of many similar relationships that can be derived from the properties of binomial coefficients [see Gradshteyn and Ryzhik (1965, Chap. 0)].

Whereas Fig. 9.7.2 is largely a grid with spacings of one-half, many $K = L - 2$ transcendentals arise from smaller units. Thus Fresnel integrals arise when the $J_0(2\sqrt{x})$ basis hypergeometric is operated on by $d^{\pm\frac{1}{4}}/dx^{\pm\frac{1}{4}}$, while Airy functions are produced by the $d^{\pm\frac{1}{3}}/dx^{\pm\frac{1}{3}}$ operators. Figure 9.7.3 shows how the fres( ) and gres( ) functions (see Section 7.6) and the fai( ) and gai( ) functions (see Section 9.1) arise in this way.

CHAPTER 10

APPLICATIONS IN THE CLASSICAL CALCULUS

This chapter will deal briefly with a few applications of the fractional calculus to problems whose formulations and solutions are normally couched in terms of integrals or derivatives alone. The number of such problems to which techniques of the fractional calculus may be successfully applied is very large—we are able only to indicate briefly the scope of this subject. As we have often found to be so, the use of differintegral operators and their general properties greatly facilitates the problem formulation and solution.

10.1 EVALUATION OF DEFINITE INTEGRALS AND INFINITE SUMS

That differintegration provides a route for the evaluation of definite integrals is illustrated here with several examples. A single example will serve to illustrate the utility of the fractional calculus as a means of summing infinite series.

Substitution of $y = x - x\lambda$ in the Riemann–Liouville definition (3.6.2) of the differintegral of x^q, where $-1 < q < 0$, gives

$$\frac{d^q x^q}{dx^q} = \frac{1}{\Gamma(-q)} \int_0^1 \frac{[1-\lambda]^q\, d\lambda}{\lambda^{q+1}} = \Gamma(q+1).$$

The further substitution $z = -\ln(\lambda)$ then leads to

$$\int_0^\infty \frac{dz}{[\exp(z) - 1]^{-q}} = \Gamma(-q)\Gamma(q+1) = \pi\csc(q\pi),$$

a result that is quite difficult to establish otherwise.

10 APPLICATIONS IN THE CLASSICAL CALCULUS

The same $y = x - x\lambda$ substitution in the Riemann–Liouville definition of the differintegral of an arbitrary function gives

$$\int_0^{x^{-q}} f(x - x\lambda)\, d(\lambda^{-q}) = \Gamma(1 - q) x^q \frac{d^q f}{dx^q}$$

for $q < 0$. Replacing λ^{-q} by z and $-1/q$ by the positive (but not necessarily integer) number p, we obtain

$$\int_0^{x^{1/p}} f(x - xz^p)\, dz = \Gamma\!\left(\frac{p+1}{p}\right) x^{-1/p} \frac{d^{-1/p} f}{dx^{-1/p}}$$

as a formula which has utility in definite integration, particularly for $x = 1$:

$$\int_0^1 f(1 - z^p)\, dz = \Gamma\!\left(\frac{p+1}{p}\right) \frac{d^{-1/p} f}{dx^{-1/p}}\bigg|_{x=1}.$$

As one example consider

$$\int_0^1 \exp(1 - z^{\frac{2}{3}})\, dz = \Gamma(\tfrac{5}{2}) \frac{d^{-\frac{3}{2}}}{dx^{-\frac{3}{2}}} \exp(x)\bigg|_{x=1}$$

$$= \tfrac{3}{4}\sqrt{\pi}\left[\exp(x)\operatorname{erf}(\sqrt{x}) - 2\sqrt{\frac{x}{\pi}}\right]_{x=1} = 1.5451,$$

and as a second

$$\int_0^1 \sin(\sqrt{1 - z^2})\, dz = \Gamma(\tfrac{3}{2}) \frac{d^{-\frac{1}{2}}}{dx^{-\frac{1}{2}}} \sin(\sqrt{x})\bigg|_{x=1}$$

$$= \tfrac{1}{2}\sqrt{\pi}[\sqrt{\pi x}\, J_1(\sqrt{x})]_{x=1} = 0.69123.$$

As we have just seen, use of the Riemann–Liouville definition of a differintegral provides an avenue for the evaluation of certain definite integrals. One might well expect that use of some of the general properties of such differintegral operators will be of even greater benefit in evaluating integrals. For example, Osler (1972c) has established the integral analog of Leibniz's rule:

(10.1.1) $$\frac{d^q[fg]}{dx^q} = \int_{-\infty}^{\infty} \binom{q}{\lambda + \gamma} \frac{d^{q-\gamma-\lambda} f}{dx^{q-\gamma-\lambda}} \frac{d^{\gamma+\lambda} g}{dx^{\gamma+\lambda}}\, d\lambda,$$

where γ is arbitrary. The choices $f = x^p$, $g = x^P$, and $\gamma = 0$ in (10.1.1) lead to

$$\int_{-\infty}^{\infty} \frac{\Gamma(q+1)\Gamma(p+1)\Gamma(P+1)\, d\lambda}{\Gamma(q-\lambda+1)\Gamma(\lambda+1)\Gamma(p-q+\lambda+1)\Gamma(P-\lambda+1)} = \frac{\Gamma(p+P+1)}{\Gamma(p+P-q+1)},$$

which is an integral extension of the identity (1.3.21). Specialization to $P = 0$, $p - q + 1 = 1$ leads to the interesting formula

$$\int_{-\infty}^{\infty} \frac{\sin(\pi\lambda)\, d\lambda}{\lambda \Gamma(\lambda + 1)\Gamma(Q - \lambda)} = \frac{\pi}{\Gamma(Q)}$$

upon setting $q + 1 = Q$. Osler (1972c) has, in fact, made use of an equation that generalizes even equation (10.1.1) to create a short table of integrals.

A similar idea of Osler's (1970a) was to use the discrete version of Leibniz's rule (see Section 5.5)

$$\frac{d^q[fg]}{dx^q} = \sum_{k=-\infty}^{\infty} \binom{q}{j+\gamma} \frac{d^{q-\gamma-k} f}{dx^{q-\gamma-k}} \frac{d^{\gamma+k} g}{dx^{\gamma+k}}$$

to derive summation formulas. Here we notice only that the choices $f = x^p$, $g = x^P$, and $\gamma = 0$ lead this time to the result

$$\sum_{k=0}^{\infty} \frac{\Gamma(q+1)\Gamma(p+1)\Gamma(P+1)}{\Gamma(q-k+1)\Gamma(k+1)\Gamma(p-q+k+1)\Gamma(P-k+1)} = \frac{\Gamma(p+P+1)}{\Gamma(p+P-q+1)}$$

which, upon selecting P to be a positive integer j, reproduces (1.3.21) exactly.

10.2 ABEL'S INTEGRAL EQUATION

The elegance and power of the fractional calculus is nicely illustrated by its use in formulating and solving the weakly singular integral equation discussed in this section, whose study by Abel in the early 19th century gave birth to the subject of integral equations. Abel was interested in the problem of the tautochrone; that is, determining a curve in the (x, y) plane such that the time required for a particle to slide down the curve to its lowest point is independent of its initial placement on the curve. More generally, one might specify the time required for descent as a function of initial height. Let us fix the lowest point of the curve at the origin and position the curve in the positive quadrant of the plane, denoting by (X, Y) the initial point and (x, y) any point intermediate between $(0, 0)$ and (X, Y) (see Fig. 10.2.1). Assuming no frictional losses, we may equate the gain in kinetic energy to the loss in potential energy:

$$\frac{m}{2}\left[\frac{d\sigma}{dt}\right]^2 = mg[Y - y],$$

where σ is the arc length along the curve measured from the origin, m is the mass of the particle, g the gravitational acceleration, and t the time. Thus, since $d\sigma/dt < 0$,

(10.2.1) $$d\sigma = -\sqrt{2g[Y - y]}\, dt.$$

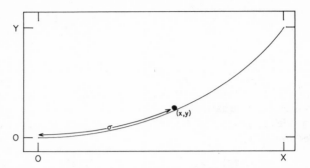

FIG. 10.2.1. The coordinate system for the tautochrone.

Now the arc length σ depends on the unknown curve. We write $\sigma = \sigma(y(t))$ to indicate the dependence of σ on the height y (which itself depends on time). Separating variables in (10.2.1) and integrating from $y = Y$ to $y = 0$ leads to

$$\sqrt{2g}\, T = \int_0^Y \frac{\sigma^{(1)}(y)\, dy}{\sqrt{Y - y}}, \tag{10.2.2}$$

where T is the time of descent. Writing $\sqrt{2g}\, T = f(Y)$ puts (10.2.2) into the form

$$f(Y) = \int_0^Y \frac{\sigma^{(1)}(y)\, dy}{\sqrt{Y - y}}, \qquad f(0) = 0, \tag{10.2.3}$$

in which we may recognize it as an integral equation of convolution type for the unknown arc length σ. Abel (1823, 1825) found the solution of equation (10.2.3) to be

$$\sigma(y) = \frac{1}{\pi} \int_0^y \frac{f(Y)\, dY}{\sqrt{y - Y}}, \tag{10.2.4}$$

essentially by Laplace transforming (10.2.3), making use of the convolution theorem, and inverting.

Of course, equation (10.2.4) still gives only an expression for the arc length $\sigma(y)$ as a function of height y along the curve in terms of the function $f(Y)$ determined by the time required for descent from an initial height Y. The relationship

$$\frac{d\sigma}{dy} = \sqrt{1 + \left[\frac{dx}{dy}\right]^2}$$

may then be used to obtain the differential equation for the curve in terms of x and y. For the simplest case when $f(Y) = \sqrt{2g}\,T$ is independent of Y one has

$$1 + \left[\frac{dx}{dy}\right]^2 = \frac{2gT^2}{\pi^2 y},$$

and the equations for the resulting tautochrone may be obtained from this as

$$x = \tfrac{1}{2}a[\theta + \sin(\theta)], \qquad y = \tfrac{1}{2}a[1 - \cos(\theta)]$$

where $a = 2gT^2/\pi^2$ and θ is the angle depicted in Fig. 10.2.2. The tautochrone is one arch of the cycloid generated by a point P on a circle of radius $\tfrac{1}{2}a$ as the circle rolls along the lower side of the line $y = a$.

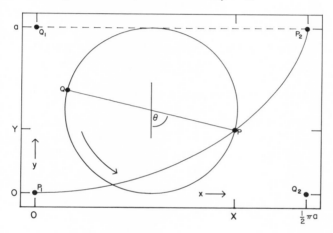

FIG. 10.2.2. The tautochrone $P_1 P_2$ is traced out by point P as the circle rolls along the line $y = a$. The point Q, initially at Q_1, ends at Q_2.

By this time the reader must surely recognize the integral in (10.2.3) as a close relative of the semiintegral of $\sigma^{(1)}$:

$$\frac{d^{-\frac{1}{2}}\sigma^{(1)}(Y)}{dY^{-\frac{1}{2}}} = \frac{1}{\Gamma(\frac{1}{2})} \int_0^Y \frac{\sigma^{(1)}(y)\,dy}{\sqrt{Y-y}}$$

so that in the language of the fractional calculus, equation (10.2.3) reads

$$f(Y) = \Gamma(\tfrac{1}{2}) \frac{d^{-\frac{1}{2}}\sigma^{(1)}}{dY^{-\frac{1}{2}}}(Y).$$

Use of the formula (5.7.13) with $q = -\tfrac{1}{2}$, $N = 1$, and Y replaced by y gives

$$f(y) = \sqrt{\pi}\left[\frac{d^{\frac{1}{2}}\sigma(y)}{dy^{\frac{1}{2}}} - \frac{y^{\frac{1}{2}}\sigma(0)}{\sqrt{\pi}}\right] = \sqrt{\pi}\,\frac{d^{\frac{1}{2}}\sigma(y)}{dy^{\frac{1}{2}}}$$

since $\sigma(0) = 0$. If we assume σ to be differintegrable, the continuity of σ at $y = 0$ may be used to rule out the presence, in σ, of a factor y^p for $p < 0$. Thus (see Section 5.7) the composition rule may be applied to give

$$\frac{d^{-\frac{1}{2}}f(y)}{dy^{-\frac{1}{2}}} = \sqrt{\pi}\,\sigma(y),$$

which is Abel's solution (10.2.4) upon writing the Riemann–Liouville form for the semiintegral. Notice how easily the inversion of the original equation (10.2.3) is performed through the use of general properties of differintegrals.

The generalized Abel equation,

$$f(Y) = \int_0^Y \frac{\sigma^{(1)}(y)\,dy}{[Y-y]^\alpha}, \qquad 0 < \alpha < 1,$$

may be inverted with equal ease through the fractional calculus.

10.3 SOLUTION OF BESSEL'S EQUATION

As an example of the way differintegration can be used to tackle classical differential equations, we here consider Bessel's equation, which arises in connection with the vibrations of a circular drumhead, as well as in other important physical applications. The modified Bessel equation, which differs only in the sign of the third term, and which arises in a number of diffusion problems, is equally amenable to the approach we here take.

The equation

$$(10.3.1) \qquad x^2 \frac{d^2 w}{dx^2} + x\frac{dw}{dx} + \left[x - \frac{v^2}{4}\right]w = 0$$

is a form of Bessel's equation. As is the rule for second-order differential equations, its general solution is a combination of two linearly independent functions w_1 and w_2 of x, each of which depends on the parameter v. The usual method of solving (10.3.1) is via an infinite series approach, but we shall demonstrate how differintegration procedures lead to a ready solution in terms of elementary functions.

We start by making either of the substitutions

$$w = x^{\pm \frac{1}{2}v} u,$$

where v denotes the nonnegative square root of v^2, so that equation (10.3.1) is transformed to

$$(10.3.2) \qquad x\frac{d^2 u}{dx^2} + [1 \pm v]\frac{du}{dx} + u = 0.$$

10.3 SOLUTION OF BESSEL'S EQUATION

We next assume that for every function u that satisfies (10.3.2) there exists a differintegrable function f, related to u by the equation

(10.3.3) $$u = \frac{d^{\frac{1}{2} \pm v} f}{dx^{\frac{1}{2} \pm v}}.$$

Moreover, use of equation (3.2.5) permits the combination of equations (10.3.2) and (10.3.3) to give

(10.3.4) $$x \frac{d^{\frac{5}{2} \pm v} f}{dx^{\frac{5}{2} \pm v}} + [1 \pm v] \frac{d^{\frac{3}{2} \pm v} f}{dx^{\frac{3}{2} \pm v}} + \frac{d^{\frac{1}{2} \pm v} f}{dx^{\frac{1}{2} \pm v}} = 0.$$

Application of Leibniz's rule allows the rewriting of equation (10.3.4) as

(10.3.5) $$\frac{d^{\frac{1}{2} \pm v} \{xf\}}{dx^{\frac{1}{2} \pm v}} - \frac{3}{2} \frac{d^{\frac{3}{2} \pm v} f}{dx^{\frac{3}{2} \pm v}} + \frac{d^{\frac{1}{2} \pm v} f}{dx^{\frac{1}{2} \pm v}} = 0$$

wherein the parameter v is no longer present as a coefficient. We next plan to decompose the operators, thus

(10.3.6) $$\frac{d^{\frac{1}{2} \pm v}}{dx^{\frac{1}{2} \pm v}} \frac{d^2 \{xf\}}{dx^2} - \frac{3}{2} \frac{d^{\frac{1}{2} \pm v}}{dx^{\frac{1}{2} \pm v}} \frac{df}{dx} + \frac{d^{\frac{1}{2} \pm v} f}{dx^{\frac{1}{2} \pm v}} = 0,$$

an equation directly convertible to

(10.3.7) $$\frac{d^2 \{xf\}}{dx^2} - \frac{3}{2} \frac{df}{dx} + f = 0$$

by the action of the $d^{-\frac{1}{2} \mp v}/dx^{-\frac{1}{2} \mp v}$ operator. Equations (10.3.6) and (10.3.5) are equivalent to each other if and only if

(10.3.8) $$[xf]_{x=0} = 0 \quad \text{and} \quad \left[\frac{d\{xf\}}{dx} \right]_{x=0} = 0$$

and

(10.3.9) $$f(0) = 0,$$

whereas (10.3.7) and (10.3.6) are equivalent if

(10.3.10) $$\frac{d^{-\frac{1}{2} \mp v}}{dx^{-\frac{1}{2} \mp v}} \frac{d^{\frac{1}{2} \pm v} g}{dx^{\frac{1}{2} \pm v}} = g \quad \text{with} \quad g = f, \frac{df}{dx}, \text{ and } \frac{d^2\{xf\}}{dx^2}.$$

Conversion of equation (10.3.7) to the canonical form

$$\frac{d^2 f}{[d(2\sqrt{x})]^2} + f = 0$$

10 APPLICATIONS IN THE CLASSICAL CALCULUS

is straightforward, whereby it follows that the two possible candidate functions f are

$$f_1 = \sin(2\sqrt{x}) \quad \text{and} \quad f_2 = \cos(2\sqrt{x}).$$

We must now inquire which, if either, of these candidate functions satisfies the requirements (10.3.8), (10.3.9), and (10.3.10), which we assumed held during our derivation. Because

$$\cos(2\sqrt{x}) = 1 - 2x + \tfrac{2}{3}x^2 - \cdots,$$

it is evident that f_2 fails to meet requirement (10.3.8) or (10.3.9) and must be rejected. However, f_1 passes these tests. The requirement

$$\frac{d^{-\frac{1}{2}+\nu}}{dx^{-\frac{1}{2}+\nu}} \frac{d^{\frac{1}{2}-\nu}g}{dx^{\frac{1}{2}-\nu}} = g \quad \text{with} \quad g = f, \ \frac{df}{dx}, \ \text{and} \ \frac{d^2\{xf\}}{dx^2},$$

one part of (10.3.10), is met by the function

$$\sin(2\sqrt{x}) = 2x^{\frac{1}{2}} - \tfrac{4}{3}x^{\frac{3}{2}} + \tfrac{4}{15}x^{\frac{5}{2}} - \cdots$$

for all values of ν (recall that we restricted ν to nonnegative values), while the other part,

$$\frac{d^{-\frac{1}{2}-\nu}}{dx^{-\frac{1}{2}-\nu}} \frac{d^{\frac{1}{2}+\nu}g}{dx^{\frac{1}{2}+\nu}} = g \quad \text{with} \quad g = f, \ \frac{df}{dx}, \ \text{and} \ \frac{d^2\{xf\}}{dx^2},$$

is met by f_1 for all ν values except the nonnegative integers.

Returning to equation (10.3.3) then, we conclude that the function

$$u_1 = \frac{d^{\frac{1}{2}-\nu}}{dx^{\frac{1}{2}-\nu}} \sin(2\sqrt{x})$$

is a solution to equation (10.3.2) for all ν values, and that

$$u_2 = \frac{d^{\frac{1}{2}+\nu}}{dx^{\frac{1}{2}+\nu}} \sin(2\sqrt{x})$$

is another solution when ν is not an integer. Our sought solutions to the original Bessel equation are thus

$$w_1(\nu, x) = x^{-\frac{1}{2}\nu} u_1 = x^{-\frac{1}{2}\nu} \frac{d^{\frac{1}{2}-\nu} \sin(2\sqrt{x})}{dx^{\frac{1}{2}-\nu}}, \quad \text{all} \ \ \nu \geq 0,$$

and

$$w_2(\nu, x) = x^{\frac{1}{2}\nu} u_2 = x^{\frac{1}{2}\nu} \frac{d^{\frac{1}{2}+\nu} \sin(2\sqrt{x})}{dx^{\frac{1}{2}+\nu}}, \quad 0 \leq \nu \neq 1, 2, \ldots.$$

The problem is now completely solved, except that a second solution is needed for integer v values. Our technique cannot reveal this second solution.

The relationship of w_1 and w_2 to the conventional notation for Bessel functions is simply

$$w_1(v, x) = \sqrt{\pi} J_{-v}(2\sqrt{x}) \quad \text{and} \quad w_2(v, x) = \sqrt{\pi} J_v(2\sqrt{x}).$$

10.4 CANDIDATE SOLUTIONS FOR DIFFERENTIAL EQUATIONS

As is the case in several facets of the fractional calculus, the composition rule lies at the crux of the derivation of the last section. Much care and labor needs to be expended in examining, at every step in an argument such as the one involved in solving Bessel's equation, the demands of this exacting rule. Such examination is especially difficult and tedious when one is dealing with unknown functions. In the present section we shall bypass all these difficulties by blithely assuming that the composition rule applies universally! We excuse this cavalier treatment of a vital theorem of the fractional calculus on the basis of the uses that will be made of the techniques of the present section.

The major task in solving a difficult ordinary differential equation is the search for candidate solutions. Having found a candidate, it is a comparatively simple matter to test whether it does or does not satisfy the original differential equation. Hence, if lack of attention to the details of the composition rule lets through a few illicit solutions, no great harm is done: These offenders will be found wanting when they are processed in an attempt to reproduce the original problem equation. Thus, in the last section, had we not so painstakingly checked the requirements of the composition rule at each stage of our derivations, the two functions

$$x^{\pm \frac{1}{2}v} \frac{d^{\frac{1}{2} \pm v} \cos(2\sqrt{x})}{dx^{\frac{1}{2} \pm v}}$$

would have emerged as possible solutions to Bessel's equation. It would, however, have been an easy step to show that these particular candidate solutions fail to regenerate Bessel's equation. On the other hand, inattention to the requirements of the composition rule may, as by excluding certain additive power terms which ought properly to have been present, allow one to miss certain correct solutions. Even then, however, the exercise may not have been a complete waste of time, since a careful examination of why a candidate solution fails to reproduce the original ordinary differential equation can give valuable clues as to how this potential solution needs to be "patched up" so as to fill the needs of the original equation.

190   10 APPLICATIONS IN THE CLASSICAL CALCULUS

With the philosophy of the foregoing in mind, let us consider the following third-order differential equation,

$$\text{(10.4.1)} \qquad p_3 \frac{d^3 w}{dx^3} + p_2 \frac{d^2 w}{dx^2} + p_1 \frac{dw}{dx} + p_0 w = 0,$$

where p_n is a polynomial in x of degree not exceeding n; that is, p_3 may be a cubic in x, p_2 a quadratic, p_1 is linear, and p_0 is necessarily a constant. In our search for candidate functions of x which hopefully will satisfy equation (10.4.1), we assume the existence of a function f of x that satisfies

$$\text{(10.4.2)} \qquad w = \frac{d^{-1-q} f}{dx^{-1-q}},$$

q being presently unspecified. Assuming the composition rule, the equation

$$\text{(10.4.3)} \qquad \frac{d^q}{dx^q}\left\{p_3 \frac{d^{2-q} f}{dx^{2-q}}\right\} + \frac{d^q}{dx^q}\left\{p_2 \frac{d^{1-q} f}{dx^{1-q}}\right\} + \frac{d^q}{dx^q}\left\{p_1 \frac{d^{-q} f}{dx^{-q}}\right\} + \frac{d^q}{dx^q}\left\{p_0 \frac{d^{-q-1} f}{dx^{-q-1}}\right\} = 0$$

arises from combining equations (10.4.1) and (10.4.2), followed by application of the d^q/dx^q operator to each term. Consider the first term in equation (10.4.3), recalling that p_3 is, in general, a cubic, say $c_3 x^3 + c_2 x^2 + c_1 x + c_0$. Application of the product rule permits the evaluation of the first term as

$$[c_3 x^3 + c_2 x^2 + c_1 x + c_0] \frac{d^2 f}{dx^2} + q[3c_3 x^2 + 2c_2 x + c_1] \frac{df}{dx}$$
$$+ q[q-1][3c_3 x + 2c_2] f + q[q-1][q-2] c_3 \frac{d^{-1} f}{dx^{-1}},$$

the composition rule having been assumed valid yet again. Similar expressions arise from each of the four terms in (10.4.3), so that the result

$$\text{(10.4.4)} \qquad \pi_3 \frac{d^2 f}{dx^2} + \pi_2 \frac{df}{dx} + \pi_1 f + \pi_0 \frac{d^{-1} f}{dx^{-1}} = 0$$

finally emerges, where π_3 is a cubic, π_2 is a quadratic, π_1 is linear, and π_0 is a constant. The constant π_0 is a composite of q and of the coefficients in the original polynomials p_3, p_2, p_1, and p_0. Since q is hitherto unrestricted we are free to assign any value to it, and it is likely that several q values (three, in general) will cause π_0 to vanish. The selection of any one such value converts equation (10.4.4) to a second-order differential equation.

10.4 CANDIDATE SOLUTIONS FOR DIFFERENTIAL EQUATIONS

Obviously, the same technique can convert a differential equation of second order to one of first order, as Liouville (1832b) demonstrated. We shall use an example of this conversion to illustrate this section. The equation

(10.4.5) $$[x^2 - 1]\frac{d^2w}{dx^2} + 2x\frac{dw}{dx} - v[v+1]w = 0$$

is Legendre's equation. To solve it by the technique just advocated, we write

$$\frac{d^q}{dx^q}\left\{[x^2-1]\frac{d^{1-q}f}{dx^{1-q}}\right\} + \frac{d^q}{dx^q}\left\{2x\frac{d^{-q}f}{dx^{-q}}\right\} - \frac{d^q}{dx^q}\left\{v[v+1]\frac{d^{-q-1}f}{dx^{-q-1}}\right\} = 0$$

and then apply Leibniz's rule. Subject to our assumption that the composition rule always holds, we get

$$[x^2-1]\frac{df}{dx} + 2[q+1]xf + [q^2+q-v^2-v]\frac{d^{-1}f}{dx^{-1}} = 0.$$

By choosing either

(10.4.6) $$q = v \quad \text{or} \quad q = -v - 1,$$

the final right-hand term vanishes leaving a first-order differential equation. The variables are separable, so that

$$\int \frac{df}{f} = 2[q+1]\int \frac{x\,dx}{1-x^2}.$$

After integration and exponentiation, the result

$$f = C[x^2 - 1]^{-1-q}$$

emerges, C being an arbitrary constant. Recalling (10.4.2) and (10.4.6), we find that

$$w_1(v, x) = \frac{d^{-1-v}[x^2-1]^{-1-v}}{dx^{-1-v}} \quad \text{and} \quad w_2(v, x) = \frac{d^v[x^2-1]^v}{dx^v}$$

appear as our two candidate solutions.

Now notice in equation (10.4.5) that replacement of v by $-v-1$ leaves the equation unchanged. It follows that solutions w_1 and w_2, which interchange on replacing v by $-v-1$, are actually identical. Our technique has, therefore, in this case, yielded a single solution. It is, however, a valid solution, being related to the conventional notation for Legendre functions by

(10.4.7) $$w_2(v, x) = \frac{d^v[x^2-1]^v}{dx^v} = 2^v\Gamma(v+1)P_v(x).$$

We may note in passing that the differintegral representation[1] for $P_\nu(x)$ embodied in (10.4.7) constitutes a generalization of the well-known Rodrigues formula (Abramowitz and Stegun, 1964, p. 334)

$$P_n(x) = \frac{1}{2^n n!} \frac{d^n[x^2 - 1]^n}{dx^n},$$

used, for integer n of course, to generate Legendre polynomials.

10.5 FUNCTION FAMILIES

Many functions are customarily defined by an integral, that is, by a $q = -1$ differintegral. By considering the behavior of the definition when q takes values other than -1, one may suitably embed the function in a family parametrized by q. We will illustrate this concept by evolving a generalized logarithm function.

The logarithm is unique in function theory in being the only transcendental function that can be generated by applying the operations of integration or differentiation to a power of $C - cx$, where C and c are constant. A collection of functions that includes a logarithm is generated by nth-order integration of $[C - cx]^{-n}$, that is, by

$$\frac{d^{-n}[C - cx]^{-n}}{dx^{-n}}, \qquad n = 1, 2, 3, \ldots,$$

where, for convenience, a lower limit of zero has been selected. The only other functions generated by

$$\frac{d^q[C - cx]^p}{dx^q}, \qquad q = 0, \pm 1, \pm 2, \ldots,$$

are algebraic and finite:

$$\frac{d^q[C - cx]^p}{dx^q} = \begin{cases} \dfrac{\Gamma(p + 1)[C - cx]^{p-q}}{\Gamma(p - q + 1)}, & q = 0, 1, 2, \ldots, \\ \dfrac{\Gamma(p + 1)}{\Gamma(p - q + 1)} \left[[C - cx]^{p-q} - \sum_{j=0}^{n-1} \binom{p + n}{j} C^{p+n-j}[-cx]^j \right], \\ \qquad\qquad\qquad\qquad\qquad p \neq q = -1, -2, \ldots, \end{cases}$$

for all p and all nonzero C and c.

---

[1] Essentially the same formula was obtained in a different context by Osler (1970a).

10.5 FUNCTION FAMILIES

The coalescence of the differintegration order with the power of the binomial is evidently the key to ensuring that the generated function family includes a logarithm. Thus, choosing $C = c = 1$, we have

$$\frac{d^{-1}[1-x]^{-1}}{dx^{-1}} = \ln\left(\frac{1}{1-x}\right),$$

$$\frac{d^{-2}[1-x]^{-2}}{dx^{-2}} = \ln\left(\frac{1}{1-x}\right) - x,$$

$$\frac{d^{-n}[1-x]^{-n}}{dx^{-n}} = \frac{1}{(n-1)!}\left[\ln\left(\frac{1}{1-x}\right) - x - \frac{x^2}{2} - \cdots - \frac{x^{n-1}}{n-1}\right].$$

Let us now define a generalized logarithm of order n, thus

$$\ln_n\left(\frac{1}{1-x}\right) = (n-1)!\frac{d^{-n}[1-x]^{-n}}{dx^{-n}}$$

for $n = 1, 2, \ldots$. Then we see that

$$\ln_1\left(\frac{1}{1-x}\right) = \ln\left(\frac{1}{1-x}\right)$$

and generally

(10.5.1) $\quad \ln_n\left(\frac{1}{1-x}\right) = \ln\left(\frac{1}{1-x}\right) - \sum_{j=1}^{n-1}\frac{x^j}{j} = \sum_{j=1}^{\infty}\frac{x^j}{j} - \sum_{j=1}^{n-1}\frac{x^j}{j} = \sum_{j=n}^{\infty}\frac{x^j}{j},$

where a Taylor expansion of $\ln(1/[1-x])$ was used. Thus the $\ln_n(\ )$ function is a "beheaded" logarithm, the first $n-1$ terms of the regular logarithmic series being absent.

We now study the implications of extending our definition to fractional v instances. From the general definition we find

$$\ln_v\left(\frac{1}{1-x}\right) = \Gamma(v)\frac{d^{-v}[1-x]^{-v}}{dx^{-v}} = \frac{d^{-v}}{dx^{-v}}\left[x\frac{v-1}{0}\right]$$

$$= x^v\left[x\frac{v-1}{v}\right] = \sum_{j=0}^{\infty}\frac{x^{j+v}}{j+v}$$

for all v. When we notice that equation (10.5.1) can equally well be written

$$\ln_n\left(\frac{1}{1-x}\right) = \sum_{j=0}^{\infty}\frac{x^{j+n}}{j+n},$$

the generalization is seen as perfect from the series expansion standpoint.

The three representations

(10.5.2) $\quad \Gamma(v)\dfrac{d^{-v}[1-x]^{-v}}{dx^{-v}}, \qquad x^v\left[x\dfrac{v-1}{v}\right], \qquad \text{and} \qquad \sum_{j=0}^{\infty}\dfrac{x^{j+v}}{j+v}$

are equivalent, except for $v = 0, -1, -2, \ldots$ and any one may be used to represent $\ln_v(1/[1-x])$. Because of the recursion

$$\ln_{v+1}\left(\frac{1}{1-x}\right) = \ln_v\left(\frac{1}{1-x}\right) - \frac{x^v}{v},$$

if the properties of $\ln_v(\ )$ are known on the interval $0 < v \leq 1$, they are known everywhere. For the important $v = \frac{1}{2}$ case, we have

$$\ln_{\frac{1}{2}}\left(\frac{1}{1-x}\right) = 2\operatorname{arctanh}(\sqrt{x}) = \ln\left(\frac{1+\sqrt{x}}{1-\sqrt{x}}\right)$$

so that a half-order logarithm is merely an ordinary logarithm with changed argument. Generalized logarithms of order $\frac{1}{4}$ and $\frac{3}{4}$ are also expressible; thus

$$\ln_{\frac{1}{4}}\left(\frac{1}{1-x}\right) = 2\operatorname{arctanh}(x^{\frac{1}{4}}) - 2\arctan(x^{\frac{1}{4}}),$$

$$\ln_{\frac{3}{4}}\left(\frac{1}{1-x}\right) = 2\operatorname{arctanh}(x^{\frac{1}{4}}) + 2\arctan(x^{\frac{1}{4}}),$$

in terms of more familiar functions.

For $v > 0$ the function $\ln_v(1/[1-x])$ is (for x real and greater than zero) single-valued, real, and finite. The corresponding function $\ln_v(1/[1+x])$, however, enters the complex plane, even for real x, unless v is an integer. This is clear from consideration of a change in the sign of x in any one of the three representations (10.5.2). In each, there is a new factor $[-1]^v$ introduced:

$$\ln_v\left(\frac{1}{1+x}\right) = \Gamma(v) \frac{d^{-v}[1+x]^{-v}}{[d(-x)]^{-v}} = (-x)^v\left[-x\frac{v-1}{v}\right] = \sum_{j=0}^{\infty} \frac{[-x]^{j+v}}{j+v}.$$

Interpreting this factor as $\cos(v\pi) + i\sin(v\pi)$, we see that for $v = \frac{1}{2}$ or $n + \frac{1}{2}$, the modified logarithm of $[1+x]^{-1}$ is purely imaginary. For example,

$$\ln_{\frac{1}{2}}\left(\frac{1}{1+x}\right) = i\sqrt{x}\left[-x\frac{-\frac{1}{2}}{\frac{1}{2}}\right] = 2i\arctan(\sqrt{x}).$$

Such behavior need not surprise us, for as simple a function as x^v behaves similarly.

The exercise we have just concluded generalized a single function into a continuum of functions. It is also possible to use the concept of differintegration to generalize a set of functions that are parametrized by a set of integers. Thus, the repeated integrals of the error function complement,

(10.5.3) $$\frac{\sqrt{\pi}}{2} i^n\operatorname{erfc}(x) = \int_x^{\infty} \frac{[y-x]^n}{n!} \exp(-y^2)\,dy,$$

parametrized by the integer n, are generalized by the differintegral definition

(10.5.4) $$\frac{\sqrt{\pi}}{2} i^v \text{erfc}\left(\frac{1}{x}\right) = x^{-v} \frac{d^{-v-1}}{dx^{-v-1}} \left\{\frac{\exp(-x^{-2})}{x^{v+2}}\right\}$$

into a continuum of functions. By a suitable change of variable, it may readily be demonstrated that definition (10.5.4) reduces to (10.5.3) when v equals the integer n.

CHAPTER 11

APPLICATIONS TO DIFFUSION PROBLEMS

Our aim in this chapter is to expose the interesting role played by differintegrals (specifically, semiderivatives and semiintegrals) in solving certain diffusion problems. Along with the wave equation and Laplace's equation, the diffusion equation is one of the three fundamental partial differential equations of mathematical physics. The books of Crank (1956), Barrer (1941), and Jost (1952), as well as any treatise on partial differential equations, provide ample background reading about this important equation which plays such a paramount role in the theories of heat conduction (Carslaw and Jaeger, 1947), diffusion (Babbitt, 1950; Crank, 1956) viscous flow (Moore, 1964), neutron migration (Davison, 1957), flow through porous media (Muskat, 1937), electrical transmission lines (Johnson, 1950), and in other instances of transport theory.

We shall not discuss conventional solutions of the diffusion equation at all. These range from closed form solutions for very simple model problems to computer methods for approximating the concentration of the diffusing substance on a network of points. Such solutions are described extensively in the literature. Our purpose, rather, is to expose a technique for partially solving a family of diffusion problems, a technique that leads to a compact equation which is first order spatially and half order temporally. We shall show that, for semiinfinite systems initially at equilibrium, our semidifferential equation leads to a relationship between the intensive variable and the flux at the boundary. Use of this relationship then obviates the need to solve the original diffusion equation in those problems for which this behavior at the boundary is of primary importance.

Each section of this chapter will first discuss some theoretical aspect of the link between transport processes and the fractional calculus, and then apply this theory to a problem of practical importance. Each problem is drawn from a different area of application to illustrate the diversity of situations amenable to our technique.

11 APPLICATIONS TO DIFFUSION PROBLEMS

As is inevitable when results proved generally are put to practice, some of the restrictions that were found to be useful in establishing the theory may be difficult, even impossible, to verify in practice. No good scientist, however, would let this prevent him from applying the theory. Indeed, applications of the theory are frequently made without such verification, and the results obtained often point the way to extensions and improvements of the previous theoretical foundations. We shall, therefore, not apologize for our inability to show that every function we shall ever want to differintegrate is a differintegrable series according to the definition given in Section 3.1. Nor will we apologize for our present inability to extend to a broader class of functions general results proved earlier in the book. We shall, in fact, freely make use of the general properties established for differintegral operators as if all our functions were differintegrable.

11.1 TRANSPORT IN A SEMIINFINITE MEDIUM

We begin our study of transport processes by considering the diffusion equation

(11.1.1) $$\frac{\partial}{\partial t} F(\xi, \eta, \zeta, t) = \kappa \nabla^2 F(\xi, \eta, \zeta, t)$$

in which F is some intensive scalar quantity (temperature, concentration, vorticity, electrical potential, or the like) that varies with time and from point to point in a three dimensional homogeneous medium, while κ is a constant appropriate to the medium and the type of transport. In equation (11.1.1), $\partial/\partial t$ effects partial differentiation with respect to time and ∇^2 is the Laplacian operator with respect to the spatial coordinates ξ, η, and ζ. When equation (11.1.1) is augmented by an initial condition and appropriate boundary conditions, its solution F is uniquely specified as a function of time and space.

Three commonly encountered boundary geometries allow a reduction from three to one in the number of spatial coordinates needed to describe transport through the medium. We shall use three different values of a geometric factor g to characterize these simplifying geometries, namely:

the convex sphere, $g = 1$;
the convex cylinder, $g = \frac{1}{2}$; and
the plane, $g = 0$.

The significance of the adjective "convex" will be clarified by glancing at item I of Fig. 11.1.1 and noting that the boundary appears convex as viewed from the diffusion medium. The term "semiinfinite" is commonly applied

11.1 TRANSPORT IN A SEMIINFINITE MEDIUM

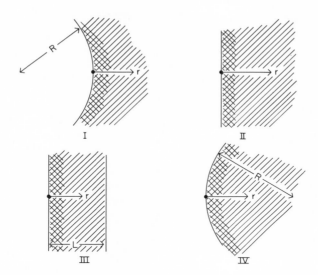

FIG. 11.1.1. This diagram illustrates, by means of cross sections, those geometries for which a single distance coordinate r suffices. Diagram I illustrates the convex spherical and cylindrical cases, $g = 1$ and $\frac{1}{2}$, it being, of course, impossible to distinguish between a sphere and a cylinder in cross section. Diagram II shows the planar, $g = 0$, case. Whereas the I and II geometries are truly semiinfinite, those diagrammed as III and IV are finite, though they behave as if they were semiinfinite at short enough times. In all four illustrations, the shaded areas represent the medium within which transport is proceeding, while the cross-hatched areas depict the zone adjacent to each boundary in which a significant perturbation occurs during the time domain $0 < t < \tau$ of interest. It is because this zone is narrow compared with L or R that these geometries behave semiinfinitely. The concave cylindrical and spherical, $g = -\frac{1}{2}$ and -1, cases are illustrated as IV.

to the three cases diagrammed as I and II; by this is meant that the diffusion medium extends indefinitely in one direction from the boundary. In these geometries the Laplacian operator simplifies so that equation (11.1.1) becomes

$$(11.1.2) \qquad \frac{\partial}{\partial t} F(r, t) - \kappa \frac{\partial^2}{\partial r^2} F(r, t) - \frac{2g\kappa}{r \pm R} \frac{\partial}{\partial r} F(r, t) = 0,$$

encompassing the three values of g. Here r is the spatial coordinate directed normal to the boundary and having its origin at the boundary surface. In the cases of spherical and cylindrical geometries, the R in equation (11.1.2) represents the radius of curvature of the surface; R is without significance in the planar case.

The motive in restricting consideration to semiinfinite geometries is that thereby the diffusion medium has only one boundary of concern, the other being "at infinity." The same situation may be achieved with media that are less than infinite in extent, such as III in Fig. 11.1.1, provided that the time domain is sufficiently restricted. As long as any perturbation which starts at the $r = 0$ boundary at time zero does not approach any other boundary of the medium within times of interest,[1] the medium behaves as if it were semiinfinite. With the proviso that we shall only ever be concerned with times short enough to ensure that a large reservoir of virtually unperturbed medium always exists, we may admit two other geometries, illustrated as IV in Fig. 11.1.1, namely,

the concave cylinder, $g = -\frac{1}{2}$; and
the concave sphere, $g = -1$,

to the class we may represent using a single distance coordinate. These two geometries are decidedly not semiinfinite, but they nevertheless behave as if they were, subject to our stated proviso. Equation (11.1.2) now applies to all five geometries, the sign to be selected in the $r \pm R$ term being that of g. Many important physical systems approximate closely to one or another of these cases.

The situations we shall treat are those in which the system is initially at equilibrium, so that

(11.1.3) $\qquad F(r, t) = F_0$, a constant; $\qquad t < 0, \quad r \geq 0.$

At $t = 0$ a perturbation of the system commences by some unspecified process occurring at the boundary. During times of interest, this perturbation does not affect regions remote from the $r = 0$ boundary, so that the relationship

(11.1.4) $\qquad F(r, t) = F_0, \qquad t \leq \tau, \qquad r = \begin{cases} \infty, & g = 0, \frac{1}{2}, 1, \\ R, & g = -\frac{1}{2}, -1, \end{cases}$

applies.

Thus our problem is described by the partial differential equation (11.1.2), initial condition (11.1.3) and the asymptotic condition (11.1.4). It may be shown (Oldham and Spanier, 1972; Oldham, 1973b) that the single equation

(11.1.5) $\qquad \dfrac{\partial}{\partial r} F(r, t) + \dfrac{1}{\sqrt{\kappa}} \dfrac{\partial^{\frac{1}{2}}}{\partial t^{\frac{1}{2}}} [F(r, t) - F_0] + \dfrac{g}{r + R} [F(r, t) - F_0] = 0$

describes the problem equally well. Equation (11.1.5) is exact in the $g = 1$ or 0 cases, and represents a short time approximation for $g = \frac{1}{2}$, $-\frac{1}{2}$, or -1. The derivation of equation (11.1.5) is given in the next section for the planar $g = 0$ case.

---

[1] Quantitatively, the requirement is that L, the dimension of the quasi-semiinfinite medium, must be related to the maximum time τ of interest by the inequality $\tau \ll L^2/\kappa$.

11.2 PLANAR GEOMETRY

In this section we will demonstrate that the relationship

$$(11.2.1) \qquad \frac{\partial}{\partial r} F(r, t) = -\frac{1}{\sqrt{\kappa}} \frac{\partial^{1/2}}{\partial t^{1/2}} F(r, t) + \frac{F_0}{\sqrt{\pi \kappa t}}$$

is a direct consequence of equations (11.1.2)–(11.1.4) in the planar $g = 0$ case, and will show how this relationship proves useful in studying heat transfer. We notice that, whereas the diffusion equation (11.1.2) is second order in space and first order in time, equation (11.2.1) [or the more general equation (11.1.5)] is first order in space and half order in time. This reduction in order has come about by incorporating the initial and asymptotic boundary conditions so that equation (11.2.1) represents the entire boundary value problem save only the boundary condition imposed at the $r = 0$ surface. This partial reduction of the complexity of the original problem will lead to important savings in a variety of situations.

The key to the derivation of equation (11.2.1) will be an understanding of the relationship between the Laplace transform of semiderivatives and semiintegrals and the Laplace transform of the undifferintegrated function. More precisely [see equation (8.1.3)], the equations[2]

$$(11.2.2) \qquad \int_0^\infty \exp(-st) \frac{d^{\pm 1/2}}{dt^{\pm 1/2}} f(t)\, dt = s^{\pm 1/2} \int_0^\infty \exp(-st) f(t)\, dt$$

will play a vital role in this derivation.

First we change variables to $u = r/\sqrt{\kappa}$ and $V(u, t) = F(r, t) - F_0$, obtaining the following equations in the new variables in the planar $g = 0$ case:

$$(11.2.3) \qquad \frac{\partial^2}{\partial u^2} V(u, t) = \frac{\partial}{\partial t} V(u, t),$$

$$(11.2.4) \qquad V(u, 0) = 0,$$

$$(11.2.5) \qquad V(\infty, t) = 0.$$

---

[2] Equation (8.1.3) for $q = \frac{1}{2}$ actually is

$$\int_0^\infty \exp(-st) \frac{d^{1/2}}{dt^{1/2}} f(t)\, dt = \sqrt{s} \int_0^\infty \exp(-st) f(t)\, dt - \frac{d^{-1/2} f}{dt^{-1/2}}(0).$$

However, the semiintegral on the right will vanish at zero provided f is differintegrable and bounded (an even milder assumption will do). We cheerfully make such an assumption here inasmuch as the physically defined functions we shall encounter throughout this chapter are necessarily finite everywhere.

11 APPLICATIONS TO DIFFUSION PROBLEMS

Upon Laplace transformation of equations (11.2.3) and (11.2.5), the relationships

(11.2.6) $$\frac{\partial^2}{\partial u^2} \overline{V}(u, s) = s\overline{V}(u, s) - V(u, 0)$$

and

(11.2.7) $$\overline{V}(\infty, s) = 0$$

are obtained, where $\overline{V}(u, s)$ is the transform of $V(u, t)$. Utilizing (11.2.4) and temporarily treating the dummy variable s as a constant, we obtain

$$\frac{d^2}{du^2} \overline{V}(u, s) - s\overline{V}(u, s) = 0$$

from equation (11.2.6). Standard methods for solving ordinary differential equations [for example, see Murphy (1960)] may be invoked to give

$$\overline{V}(u, s) = P_1(s) \exp(u\sqrt{s}) + P(s) \exp(-u\sqrt{s}),$$

where P_1 and P are arbitrary functions of s. The boundary condition (11.2.7) shows that $P_1(s) = 0$ so that

(11.2.8) $$\overline{V}(u, s) = P(s) \exp(-u\sqrt{s}),$$

where $P(s)$ remains unspecified.

The unknown function $P(s)$ depends, of course, upon the boundary condition at $r = 0$. Leaving this unspecified for now, we may eliminate $P(s)$ between (11.2.8) and the expression

$$\frac{\partial}{\partial u} \overline{V}(u, s) = -\sqrt{s} P(s) \exp(-u\sqrt{s}),$$

which results upon differentiating (11.2.8). The resulting equation in transform space,

$$\frac{\partial}{\partial u} \overline{V}(u, s) = -\sqrt{s}\, \overline{V}(u, s),$$

may be inverted, utilizing equation (11.2.2) to give

$$\frac{\partial}{\partial u} V(u, t) = -\frac{\partial^{\frac{1}{2}}}{\partial t^{\frac{1}{2}}} V(u, t).$$

Restoration of the original variables produces (11.2.1).

The exploitation of equation (11.2.1) will be illustrated with an example drawn from the theory of heat conduction. Specifically, consider the problem of determining the heat flux at the surface of a heat conductor from measurements of the surface temperature. This need arises in studies of heat transfer in wind tunnels (Meyer, 1960; Allegre, 1970) and other engineering applications.

The equation of one-dimensional heat conduction in a semiinfinite planar medium is

$$(11.2.9) \qquad \frac{\partial}{\partial t} T(r, t) = \frac{K}{\rho\sigma} \frac{\partial^2}{\partial r^2} T(r, t), \qquad 0 \leq r \leq \infty, \quad 0 \leq t \leq \infty,$$

where K, ρ, and σ are respectively the conductivity, density, and specific heat of the conducting material, T is the difference between the local and the ambient temperature, t the time, and r the distance from the surface of interest. Upon identification of T with F and κ with $K/\rho\sigma$, equations (11.2.9) and (11.1.2) are seen to be identical in the planar $g = 0$ case. Appropriate boundary conditions for this problem, analogous to equations (11.1.3) and (11.1.4), are

$$T(r, 0) = 0 \quad \text{and} \quad T(\infty, t) = 0.$$

The sought surface heat flux is

$$J(t) \equiv -K \frac{\partial}{\partial r} T(0, t),$$

which Meyer (1960) showed to be obtainable in the form

$$(11.2.10) \qquad J(t) = \sqrt{\frac{K\rho\sigma}{4\pi}} \left[\frac{2T(0, t)}{\sqrt{t}} + \int_0^t \frac{T(0, t) - T(0, \tau)}{[t - \tau]^{3/2}} d\tau \right].$$

Our approach to this problem proceeds directly from equation (11.2.1) (replacing F by T, κ by $K/\rho\sigma$, specializing to $r = 0$, and remembering that T_0 is zero) as follows:

$$\sqrt{\frac{\rho\sigma}{K}} \frac{\partial^{1/2}}{\partial t^{1/2}} T(0, t) = -\frac{\partial}{\partial r} T(0, t) = \frac{J(t)}{K}.$$

Thus,

$$J(t) = \sqrt{K\rho\sigma} \frac{d^{1/2}}{dt^{1/2}} T(0, t),$$

the desired heat flux being obtainable from the surface temperature by simple semidifferentiation. Reference to the table in Section 7.1 shows that our result and Meyer's equation (11.2.10) are equivalent.

11.3 SPHERICAL GEOMETRY

The derivation in the $g = 1$ case of equation (11.1.5),

(11.3.1) $$\frac{\partial}{\partial r} F(r, t) = \frac{-1}{\sqrt{\kappa}} \frac{\partial^{\frac{1}{2}}}{\partial t^{\frac{1}{2}}} [F(r, t) - F_0] - \frac{F(r, t) - F_0}{r + R},$$

follows a course so similar to that employed in the previous section that we shall omit it, referring the interested reader to Oldham and Spanier (1972) and Oldham (1973b). We shall demonstrate the utility of equation (11.3.1) with an example drawn from electrochemistry.

Electroanalytical chemists frequently employ a hanging mercury sphere (radius R) as an electrode immersed in an aqueous solution. By negatively polarizing the electrode it is possible to cause the surface electroreduction of any reducible species (e.g., metal ions, dissolved oxygen) that might be present in the solution at low concentrations, and thereby effect a chemical analysis. One such method, semiintegral electroanalysis (Grenness and Oldham, 1972, and Oldham, 1973a) utilizes equation (11.3.1).

Imagine that initially the electroreducible species is present throughout the solution at a uniform concentration C_0 and that its transport through the solution is solely by diffusion with a diffusion coefficient D. The electrode is then given progressively more negative electrical potential, leading to a monotonic diminution of the surface concentration $C(0, t)$ of the electroreducible species. This, in turn, leads to the diffusion of the species towards the electrode in accordance with Fick's first law,

(11.3.2) $$-J(0, t) = D \frac{d}{dr} C(0, t).$$

Moreover, Faraday's electrochemical law asserts the proportionality

(11.3.3) $$i(t) = -4\pi R^2 n F_y J(0, t)$$

between the surface flux[3] $J(0, t)$ and the electric current $i(t)$ that flows from the electrode to the external circuit. The constants in equation (11.3.3) are the electrode area $4\pi R^2$, the number n of electrons needed to reduce each molecule of diffusant, and Faraday's constant F_y ($= 96,500$ coulomb/equivalent). When equation (11.3.1) is specialized to $r = 0$ and its symbols replaced by those appropriate to the present problem,

$$\frac{d}{dr} C(0, t) = \frac{-1}{\sqrt{D}} \frac{d^{\frac{1}{2}}}{dt^{\frac{1}{2}}} \{C(0, t) - C_0\} - \frac{C(0, t) - C_0}{R}.$$

[3] Sign conventions demand that the flux be negative when transport occurs *toward* the coordinate origin from positive values of r.

emerges. Combination with equations (11.3.2) and (11.3.3) now leads to the expression

$$(11.3.4) \quad i(t) = -4\pi R^2 n F_y \sqrt{D} \left[\frac{d^{\frac{1}{2}}}{dt^{\frac{1}{2}}} \{C(0,t) - C_0\} + \frac{\sqrt{D}}{R} [C(0,t) - C_0] \right],$$

relating the time-dependent current to the surface concentration $C(0,t)$. Semiintegration of equation (11.3.4) and application of the composition rule, valid for bounded $C(0,t)$ (see footnote 2), yields

$$(11.3.5) \quad \frac{d^{-\frac{1}{2}}}{dt^{-\frac{1}{2}}} i(t) = K C_0 \left[\left\{ 1 - \frac{C(0,t)}{C_0} \right\} + \frac{\sqrt{D}}{R} \frac{d^{-\frac{1}{2}}}{dt^{-\frac{1}{2}}} \left\{ \frac{C_0 - C(0,t)}{C_0} \right\} \right]$$

after rearrangement and after adoption of K as a convenient abbreviation for the constant $4\pi R^2 n F_y \sqrt{D}$.

If the electrode polarization becomes progressively greater, a time $t = \tau$ will be reached after which $C(0,t)$ is virtually zero. Equation (11.3.5) then becomes

$$(11.3.6) \quad \frac{d^{-\frac{1}{2}}}{dt^{-\frac{1}{2}}} i(t \geqq \tau) = K C_0 \left\{ 1 + \frac{\sqrt{D}}{R} \left[\frac{d^{-\frac{1}{2}}}{dt^{-\frac{1}{2}}} \left\{ \frac{C_0 - C(0,t)}{C_0} \right\} \right]_{t \geqq \tau} \right\}.$$

Were it not for the curvature correction term

$$\frac{\sqrt{D}}{R} \left[\frac{d^{-\frac{1}{2}}}{dt^{-\frac{1}{2}}} \left\{ \frac{C_0 - C(0,t)}{C_0} \right\} \right]_{t \geqq \tau},$$

equation (11.3.6) would be a simple result asserting that the semiintegral of the current becomes a constant for $t \geqq \tau$, which constant is proportional to the initial concentration C_0. Hence measuring $[d^{-\frac{1}{2}} i/dt^{-\frac{1}{2}}]_{t \geqq \tau}$ would permit quantitative chemical analysis of the reducible species, were it not for the curvature correction term.

For a typical electrochemical experiment ($R \approx 10^{-3}$ m, $\tau \approx 1$ sec, $D \approx 10^{-9}$ m^2/sec), the curvature correction term, necessarily less[4] than $[2/R]\sqrt{D\tau/\pi}$ at $t = \tau$, is small compared with unity and is frequently ignored. Alternatively, the following approximation scheme may be adopted (see Fig. 11.3.1 for clarification):

$$\frac{\sqrt{D}}{R} \left[\frac{d^{-\frac{1}{2}}}{dt^{-\frac{1}{2}}} \left\{ \frac{C_0 - C(0,t)}{C_0} \right\} \right]_{t \geqq \tau} \approx \frac{\sqrt{D}}{R} \left[\frac{d^{-\frac{1}{2}}}{dt^{-\frac{1}{2}}} H(t - t_{\frac{1}{2}}) \right]_{t \geqq \tau}$$

$$= \left[\frac{2}{R} \sqrt{\frac{D[t - t_{\frac{1}{2}}]}{\pi}} \right]_{t \geqq \tau},$$

---

[4] That this is so may be seen by replacing $[C_0 - C(0,t)]/C_0$ by its upper bound, unity, in equation (11.3.6).

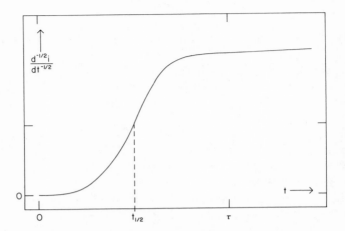

FIG. 11.3.1. The curve shows how the semiintegral of the faradaic current increases sigmoidally with time to become almost constant after $t = \tau$. The slight inconstancy is caused by electrode curvature. In an approximate procedure for correcting for this effect, the sigmoid curve is replaced by a Heaviside step function at $t = t_{\frac{1}{2}}$, the time at which the semiintegral reaches half its $t = \tau$ value.

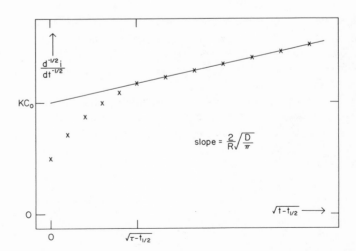

FIG. 11.3.2. To correct for the curvature effect, the semiintegral of the current is plotted versus $\sqrt{t - t_{\frac{1}{2}}}$ and extrapolated to find KC_0. The points are linear only for times exceeding τ, corresponding to the semintegral's having attained its limiting value.

11.4 INCORPORATION OF SOURCES AND SINKS

where H is the Heaviside unit function and $t_{\frac{1}{2}}$ is the time at which $C(0, t) \approx \frac{1}{2}C_0$. The graphical procedure indicated in Fig. 11.3.2 then permits a correction to be made for the small curvature effect.

11.4 INCORPORATION OF SOURCES AND SINKS

Equation (11.1.2) describes the diffusion of an entity in the absence of sources and sinks; that is, where there is no creation or annihilation of the diffusing substance within the medium. Many practical problems, however, require the inclusion of volume sources or sinks. We now turn our attention to this subject.

The equation

(11.4.1) $$\frac{\partial}{\partial t} F(r, t) - \kappa \frac{\partial^2}{\partial r^2} F(r, t) = S - kF(r, t)$$

replaces the $g = 0$ instance of equation (11.1.2) when the transport is accompanied by a constant source S and a first-order removal process, embodied in the $kF(r, t)$ term. We shall assume that the uniform steady-state condition

(11.4.2) $$F(r, t) = S/k, \quad t \leq 0,$$

is in effect prior to the time $t = 0$. We shall sketch a derivation of the relationship

(11.4.3) $$F(0, t) = \frac{S}{k} - \sqrt{\kappa} \exp(-kt) \frac{d^{-\frac{1}{2}}}{dt^{-\frac{1}{2}}} \left[\exp(kt) \frac{d}{dr} F(0, t) \right]$$

between the boundary value $F(0, t)$ of the intensive variable and the boundary value of the flux, which is proportional to the term $-dF(0, t)/dr$.

Upon Laplace transformation of equation (11.4.1), we obtain the equation

$$s\bar{F}(r, s) - F(r, 0) - \kappa \frac{\partial^2}{\partial r^2} \bar{F}(r, s) = \frac{S}{s} - k\bar{F}(r, s).$$

Use of our initial condition (11.4.2) leads to the ordinary differential equation

(11.4.4) $$\frac{d^2}{dr^2} \bar{F}(r, s) - \frac{s + k}{\kappa} \bar{F}(r, s) + \frac{S}{\kappa} \left[\frac{s + k}{sk} \right] = 0$$

in the transform of F. The most general solution of equation (11.4.4) is

(11.4.5) $$\bar{F}(r, s) - \frac{S}{sk} = P(s) \exp\left(-r\sqrt{\frac{s+k}{\kappa}}\right) + P_1(s) \exp\left(r\sqrt{\frac{s+k}{\kappa}}\right),$$

where $P(s)$ and $P_1(s)$ are arbitrary functions of s. If the geometry is semi-infinite, which we now assume, the physical requirement that F remain bounded as r tends to infinity demands that $P_1(s) \equiv 0$. Eliminating the function $P(s)$ between equation (11.4.5) and the equation obtained from it upon differentiation with respect to r, we find

$$\bar{F}(r, s) = \frac{S}{sk} - \sqrt{\frac{\kappa}{s+k}} \frac{d}{dr} \bar{F}(r, s).$$

It now requires only Laplace inversion (see Section 8.1) and specialization to $r = 0$ to obtain equation (11.4.3).

An interesting application of the preceding theory arises in modeling diffusion of atmospheric pollutants. We now describe such a model problem.

Consider a vertical column of unstirred air into the base of which a pollutant commences to be injected at time $t = 0$. Prior to $t = 0$ the air is unpolluted, so that

$$C(r, 0) = 0,$$

where $C(r, t)$ denotes the pollutant concentration at height r at time t. After $t = 0$ the rate of pollutant injection is some unspecified function $J(t)$ of time. The pollutant reacts with air by some chemical reaction that is first order (or pseudo-first order) with a rate constant k. The pollutant diffuses through air with a diffusion coefficient D that is assumed independent of height and of concentration. There are no volume sources of pollutants. We seek to relate the ground-level pollutant concentration

$$C(t) \equiv C(0, t)$$

to the input flux $J(t)$.

Identifying $C(0, t)$ with $F(0, t)$ and D with κ in equation (11.4.3), and setting $S = 0$ and $J(0, t) = -D \, dC(0, t)/dr$, we discover that

(11.4.6) $$C(t) = \frac{\exp(-kt)}{\sqrt{D}} \frac{d^{-\frac{1}{2}}}{dt^{-\frac{1}{2}}} \{\exp(kt)J(t)\}.$$

The generality of equation (11.4.6) is worth emphasizing. It enables the ground concentration of pollutant to be predicted from the rate of pollutant generation for *any* time dependent $J(t)$. Similarly, the inverse of equation (11.4.6),

(11.4.7) $$J(t) = \sqrt{D} \exp(-kt) \frac{d^{\frac{1}{2}}}{dt^{\frac{1}{2}}} \{\exp(kt)C(t)\},$$

permits the generation rate to be reconstructed from a record of the time variation of ground pollution levels.

11.4 INCORPORATION OF SOURCES AND SINKS

As a trivially simple application of equation (11.4.6), consider the case where the generation rate $J(t)$, zero prior to $t = 0$, is a constant J thereafter. Then

$$C(t) = \frac{\exp(kt)}{\sqrt{D}} \frac{d^{-\frac{1}{2}}}{dt^{-\frac{1}{2}}} \{J \exp(kt)\} = \frac{J \operatorname{erf}(\sqrt{kt})}{\sqrt{Dk}},$$

which shows that the pollutant concentration will rise to reach a final constant level of $J/\sqrt{Dk}$ and that the time to reach one-half of this final level is $0.23/k$.

As a more realistic example, consider the pollution generation rate to be sinusoidal with a mean value J and a minimum value of zero:

(11.4.8) $$J(t) = J + J \sin(at).$$

With a equal to $\pi/[12 \text{ hours}]$ and $t = 0$ corresponding to 9.00 A.M., this could represent a diurnal variation in pollution generation, typified by automobile traffic. Introduction of equation (11.4.8) into (11.4.6) followed by the indicated semiintegration leads eventually to the result

(11.4.9)
$$C(t) = \frac{J \operatorname{erf}(\sqrt{kt})}{\sqrt{Dk}} - \frac{J \exp(-kt)}{\sqrt{D[k^2 + a^2]}} \{\beta \operatorname{Rew}([\beta + i\alpha]\sqrt{t}) + \alpha \operatorname{Imw}([\beta + i\alpha]\sqrt{t})\}$$
$$+ \frac{J}{\sqrt{D[k^2 + a^2]}} \{\beta \cos(at) - \alpha \sin(at)\},$$

where α and β are positive quantities defined by

$$2\alpha^2 = \sqrt{k^2 + a^2} + k, \qquad 2\beta^2 = \sqrt{k^2 + a^2} - k,$$

and Rew( ) and Imw( ) are the real and imaginary parts of the complex error function of Faddeeva and Terent'ev (1961).

As t becomes large, the transient terms within equation (11.4.9) vanish and leave

$$C(t \to \infty) = \frac{J}{\sqrt{D}} \left[\frac{1}{\sqrt{k}} + \frac{1}{[k^2 + a^2]^{\frac{1}{4}}} \sin\left(at - \arctan\left(\frac{\beta}{\alpha}\right)\right) \right].$$

This relationship gives

$$\frac{J}{\sqrt{D}} \left[\frac{1}{\sqrt{k}} \pm \frac{1}{[k^2 + a^2]^{\frac{1}{4}}} \right]$$

as the extreme pollutant concentrations, with the peak level occurring at some time between 3.00 and 6.00 P.M.

11.5 TRANSPORT IN FINITE MEDIA

Our equation (11.2.1) was derived on the basis that the transport medium was semiinfinite in extent, $0 \leq r < \infty$. The modification engendered when r has a finite upper bound L is the subject of the present section. Three cases of the boundary will be treated: first as a perfect sink

$$F(L, t) = 0,$$

second as a boundary of zero flux

$$\frac{d}{dr} F(L, t) = 0,$$

and third as a plane in which the proportionality

$$\frac{d}{dr} F(L, t) \propto F(L, t)$$

is enforced. As we shall see, the finiteness of L destroys the simplicity of the formulation in terms of the fractional calculus. Our interest will concentrate on the short-time approximation to the finite problem; that is, we shall study the breakdown of equation (11.2.1) as L becomes discernibly less than infinity.

The physical setting will be that of electrical conduction along a transmission line,[5] and we first establish that equation (11.2.1) does, in fact, hold for an infinite line.

A two-conductor transmission line, exemplified by a coaxial cable, generally has a number of impedances associated with it, including inductance along and between the conductors and leakage conductance between them (Johnson, 1950). Here, however, we restrict consideration to an idealized transmission line in which resistance along one conductor and capacitance between the two conductors are the only significant elements. Such a line is symbolized in Fig. 11.5.1. By ρ and γ we denote the resistance and capacitance per unit length of the line.

The potential and current at the end of the resistive–capacitative line will be represented by $e(0, t)$ and $i(0, t)$. Similarly, $e(r, t)$ and $i(r, t)$ will be used to represent the interconductor potential and intraconductor currents at a distance r from the end of the line at time t. These definitions will be clarified by reference to Fig. 11.5.1.

---

[5] A very similar circumstance, though with added complexity arising from source terms, is to be found in models of the propagation of nerve impulses.

11.5 TRANSPORT IN FINITE MEDIA

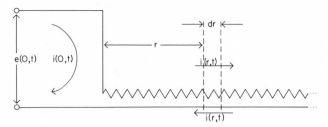

FIG. 11.5.1. Diagram displaying the symbolism adopted in the discussion of resistive–capacitative transmission lines. The interconductor potential is $e(r, t)$ at distance r.

The line is initially at rest, the potential and current being everywhere zero at $t = 0$. Subsequently a signal is applied to the end of the line, as a result of which a perturbation is transmitted down the line. The equations governing the transmission may be deduced by applying Ohm's law (the voltage difference developed by the passage of a current i through a resistor equals the product iR of the current and the resistance R of the resistor)

$$e(r + dr, t) = e(r, t) - [\rho \, dr]i(r, t)$$

to a section of line of length dr, and relating the interconductor reactance current, by means of Coulomb's law (the current flowing through a capacitor equals the product $C \, dE/dt$ of the capacitance C of the capacitor and the time derivative of the potential E across it)

$$i(r, t) - i(r + dr, t) = [\gamma \, dr]\frac{\partial}{\partial t} e(r, t),$$

to the time derivative of the interconductor potential. These two equations may be rewritten as

(11.5.1) $$\frac{\partial}{\partial r} e(r, t) + \rho i(r, t) = 0$$

and

$$\frac{\partial}{\partial r} i(r, t) + \gamma \frac{\partial}{\partial t} e(r, t) = 0,$$

and combined into the single partial differential equation

(11.5.2) $$\frac{\partial^2}{\partial r^2} e(r, t) = \rho\gamma \frac{\partial}{\partial t} e(r, t)$$

by elimination of the current.

Comparison with equation (11.1.2) shows (11.5.2) to be a typical one-dimensional diffusion equation in which $1/\rho\gamma$ plays the role of κ. The procedures of Section 11.2 may be employed to convert equation (11.5.1) to

$$\frac{\partial}{\partial r} e(r, t) = -\sqrt{\rho\gamma}\, \frac{\partial^{\frac{1}{2}}}{\partial t^{\frac{1}{2}}} e(r, t) \tag{11.5.3}$$

by incorporation of the initial condition

$$e(r, t) = 0, \quad \text{all } r, \quad t \leq 0, \tag{11.5.4}$$

and the semiinfiniteness condition

$$e(\infty, t) = 0, \quad \text{all } t < \infty.$$

We may now use equation (11.5.1) and specialize to $r = 0$ to obtain

$$i(0, t) = -\frac{1}{\rho} \frac{d}{dr} e(0, t) = \sqrt{\frac{\gamma}{\rho}}\, \frac{d^{\frac{1}{2}}}{dt^{\frac{1}{2}}} e(0, t). \tag{11.5.5}$$

In words, equation (11.5.5) demonstrates that the current drawn by a resistive–capacitative transmission line of infinite length is proportional to the time-semiderivative of the applied voltage signal.

We now return to equation (11.5.2) to consider its solution when the transmission line is of finite length L. On Laplace transformation, equation (11.5.2) becomes the ordinary differential equation

$$\frac{d^2}{dr^2} \bar{e}(r, s) = \rho\gamma[s\bar{e}(r, s) - e(r, 0)],$$

of which the general solution, after incorporation of the initial condition (11.5.4), is

$$\bar{e}(r, s) = P(s) \exp(-r\sqrt{\rho\gamma s}) + P_1(s) \exp(r\sqrt{\rho\gamma s}). \tag{11.5.6}$$

As before, $P(s)$ and $P_1(s)$ are arbitrary functions of the dummy variable s. Unlike our Section 11.2 procedure, however, we do not now find that $P_1(s)$ vanishes.

If the $r = L$ end of the transmission line is short circuited, $e(L, t)$ is necessarily zero, so that

$$0 = \bar{e}(L, s) = P(s) \exp(-L\sqrt{\rho\gamma s}) + P_1(s) \exp(L\sqrt{\rho\gamma s}).$$

This equation now provides an interrelationship between $P(s)$ and $P_1(s)$ which can be used to eliminate the latter from equation (11.5.6). Following this procedure, the result

$$\bar{e}(r, s) = P(s)[\exp(-r\sqrt{\rho\gamma s}) - \exp([r - 2L]\sqrt{\rho\gamma s})] \tag{11.5.7}$$

11.5 TRANSPORT IN FINITE MEDIA

emerges. The remaining arbitrary function $P(s)$ is eliminated at this stage, as in Section 11.2, between equation (11.5.7) and its r derivative. This leads to

$$\frac{\partial}{\partial r} \bar{e}(r, s) = -\sqrt{\rho\gamma s}\, \bar{e}(r, s) \coth([L - r]\sqrt{\rho\gamma s})$$

after some algebra, and to

(11.5.8) $$\frac{d}{dr} \bar{e}(0, s) = -\sqrt{\rho\gamma s}\, \bar{e}(0, s) \coth(L\sqrt{\rho\gamma s})$$

on specialization to $r = 0$.

For large enough L, the hyperbolic cotangent term in equation (11.5.8) approximates to unity for all but the smallest values of s (corresponding to large values of t). This cotangent term lies within 1% of unity provided

(11.5.9) $$L\sqrt{\rho\gamma s} \geq 2.6.$$

Now, $L\rho$ and $L\gamma$ are, respectively, the total resistance R and total capacitance C of the finite line, so that inequality (11.5.9) may be rewritten

$$\frac{1}{s^2} \leq \frac{RC}{7.1s},$$

which gives

(11.5.10) $$t \leq 0.14 RC$$

on inversion.[6] Hence, a short-circuited finite line reproduces the properties of an infinite line to within 1%, for times up to about 14% of the RC time constant of the finite line.

The above discussion referred to a finite line the distant end of which was short circuited, as shown in I in Fig. 11.5.2. For an open-circuited termination, diagrammed as II, the analog of equation (11.5.8) is found to be

(11.5.11) $$\frac{d}{dr} \bar{e}(0, s) = -\sqrt{\rho\gamma s}\, \bar{e}(0, s) \tanh(L\sqrt{\rho\gamma s}).$$

---

[6] Strictly, one may not invert an inequality in this way. However, if one examines the departures of the exact Laplace inverse of equation (11.5.8), namely,

$$\frac{d}{dr} e(0, t) = -2\sqrt{\rho\gamma} \sum_{j=0}^{\infty} \int_0^t [t - \tau]^{-\frac{3}{2}} \exp\left(\frac{-j^2 RC}{t - \tau}\right) \frac{d^{\frac{1}{2}}}{d\tau^{\frac{1}{2}}} e(0, \tau)\, d\tau,$$

from result (11.5.5), one finds that these are very close to 1% for t values of $0.14 RC$ for a variety of simple $e(0, t)$ signals.

Since this result is strictly complementary to (11.5.8), the same time restriction, namely, inequality (11.5.10), governs the range over which the open-circuited finite line adequately (that is to within 1%) simulates a transmission line of infinite length.

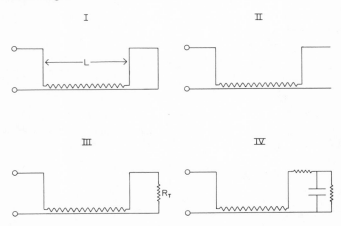

FIG. 11.5.2. Four terminations of a finite resistive–capacitative transmission line. Diagrams I and II show, respectively, a short-circuited and an open-circuited termination. In III a single resistor R_T provides a terminal impedance, while three components fill this role in diagram IV.

Judicious termination of a finite resistive–capacitative line, as for instance by a well-chosen terminating resistor R_T shown in Fig. 11.5.2 as III, will greatly increase the time range over which the transmission line behaves adequately as a semidifferentiator. To investigate the optimum value of this resistor, we need to carry out the following, rather lengthy, analysis. We first note that the Ohm's law requirement

$$e(L, t) = R_T i(L, t) = -\frac{R_T}{\rho}\frac{d}{dr}e(L, t)$$

at the $r = L$ boundary transforms to

$$\bar{e}(L, s) = -\frac{R_T}{\rho}\frac{d}{dr}\bar{e}(L, s),$$

which may be combined with equation (11.5.6) to provide the relationship

$$P_1(s) = P(s)\frac{R_T\sqrt{\gamma s} - \sqrt{\rho}}{R_T\sqrt{\gamma s} + \sqrt{\rho}}\exp(-2L\sqrt{\rho\gamma s})$$

11.5 TRANSPORT IN FINITE MEDIA

between the arbitrary functions $P_1(s)$ and $P(s)$. We now introduce this relationship into (11.5.6) to eliminate $P_1(s)$ and provide an expression for $\bar{e}(r, s)$ in terms of $P(s)$. This expression is next differentiated with respect to r and $P(s)$ is in turn eliminated between the expression and its derivative. This results in

(11.5.12)
$$\frac{d}{dr}\bar{e}(0, s) = -\sqrt{\rho\gamma s}\,\bar{e}(0, s)\frac{[R_T\sqrt{\gamma s} + \sqrt{\rho}]\exp(2L\sqrt{\rho\gamma s}) - R_T\sqrt{\gamma s} + \sqrt{\rho}}{[R_T\sqrt{\gamma s} + \sqrt{\rho}]\exp(2L\sqrt{\rho\gamma s}) + R_T\sqrt{\gamma s} - \sqrt{\rho}},$$

which, as a consideration of Fig. 11.5.2 would lead one to expect, reduces to (11.5.8) on setting $R_T = 0$ and to (11.5.11) on letting $R_T \to \infty$. In equation (11.5.12) the quotient term, which may be rewritten as

(11.5.13)
$$\frac{R_T\sqrt{Cs}\sinh(\sqrt{RCs}) + \sqrt{R}\cosh(\sqrt{RCs})}{R_T\sqrt{Cs}\cosh(\sqrt{RCs}) + \sqrt{R}\sinh(\sqrt{RCs})} \equiv 1 + \Delta,$$

defines a relative error Δ that expresses the extent to which there is a departure from the infinite length case. When s^{-1} is small, corresponding to short times, Δ is negligible; but as s^{-1} increases, Δ becomes negative, passes through a minimum at some critical value s_c of s, crosses zero, and eventually acquires

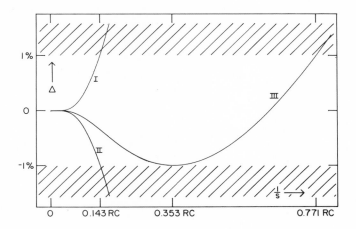

FIG. 11.5.3. Departures from perfection in semidifferentiating arising from the finiteness of a resistive–capacitative transmission line. The notations I, II, and III refer to the terminations shown in Fig. 11.5.2, the terminating resistor R_T in III having been optimally chosen for a 1% semidifferentiating error. The times $0.143RC$ and $0.771RC$ mark the limits of valid semidifferentiation for the I and II cases and for III, respectively.

ever-increasing positive values, as illustrated in Fig. 11.5.3. By differentiating Δ with respect to s and setting $d\Delta/ds$ to zero, it can be shown that the critical value of s is given by

$$\frac{1}{s_c} = \frac{R_T{}^2 C}{R + R_T}.$$

By substituting this expression back into equation (11.5.13), one can solve for values of the parameter R_T/R that cause the minimum Δ value to correspond to any chosen percentage. For a $-1\% \Delta$, for example, an R_T/R ratio of 0.797 is needed. The effect of this terminating resistor in prolonging the time interval during which the action of the finite transmission line closely approximates that of an infinite line is best brought out in Fig. 11.5.3, wherein the performances of the I, II, and III configurations of Fig. 11.5.2 are compared. Note that the time limit for III has been extended over five-fold from that given by inequality (11.5.10) for I and II to

$$t \leq 0.77 RC$$

for the optimally terminated configuration III.

Still greater improvements are possible by refining the termination through addition of a capacitor, for example, or the network shown as IV in Fig. 11.5.2, but we shall not consider such circuits further. These matters are of some practical importance in designing circuitry to effect semidifferentiation since, of course, an infinite transmission line is an abstraction. The reader will recall from Section 8.3 that discrete resistors and capacitors may be used, not only as terminating elements, but to simulate the entire transmission line in a practical semidifferentiating (or semiintegrating) network.

11.6 DIFFUSION ON A CURVED SURFACE

Whereas most diffusion problems are representable by a partial differential equation of the second order, the situation we consider in this section requires a fourth-order equation for its description. The fractional calculus may be used to solve these problems and introduces differintegrals other than those of one-half order.

Where the flat surface of a homogeneous metal is intersected by a grain boundary normal to the surface, a groove slowly develops as a result of the higher chemical potential of the metal atoms in the grain boundary plane. The metal removed in the groove formation appears as ridges parallel to the

11.6 DIFFUSION ON A CURVED SURFACE

groove as depicted in profile in Fig. 11.6.1. It may be shown (Mullins, 1957) that, on the assumption that the metal atoms move by surface diffusion under a driving force provided by the variation of surface curvature, the equation

$$\frac{\partial}{\partial t} z(x, t) = -b^2 \frac{\partial^4}{\partial x^4} z(x, t)$$

applies for small values of the angle β (see Fig. 11.6.1). Here x represents distance measured from the grain boundary parallel to the original surface. The constant b^2 is proportional to the surface diffusion coefficient of the metal. The initial condition

$$z(x, t) = 0$$

and the boundary conditions

$$z(\pm\infty, t) = \frac{\partial^2}{\partial x^2} z(\pm\infty, t) = 0,$$

specifying the semiinfiniteness of the problem, generally apply.

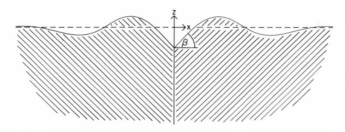

FIG. 11.6.1. An intergranular groove depicted in cross section. The dashed line denotes the original metal surface.

Robertson (1971) solved the above system of equations by employing a technique similar to that used in Section 11.2. His solution,

$$\frac{\partial}{\partial x} z(x, t) = b \frac{\partial^{-\frac{1}{2}}}{\partial t^{-\frac{1}{2}}} \left\{ \frac{\partial^3}{\partial x^3} z(x, t) \right\} - \sqrt{\frac{2}{b}} \frac{\partial^{\frac{1}{4}}}{\partial t^{\frac{1}{4}}} z(x, t),$$

relates the slope of the surface to the semiintegral of $\partial^3 z/\partial x^3$ and to the quarter-order temporal differintegral of the surface height. However, $\partial^3 z/\partial x^3$ is proportional to the local flux of metal atoms along the surface, which symmetry demands be zero at $x = 0$. By applying this boundary condition and the $d^{-\frac{1}{4}}/dt^{-\frac{1}{4}}$ operator, Robertson then obtained

$$z(0, t) = -\sqrt{\frac{b}{2}} \frac{d^{-\frac{1}{4}}}{dt^{-\frac{1}{4}}} \left\{ \frac{d}{dx} z(0, t) \right\}$$

11 APPLICATIONS TO DIFFUSION PROBLEMS

as the equation dictating the groove depth as a function of time. If the angle β is treated as a constant, we find

$$z(0, t) = -\sqrt{\frac{b}{2}} \frac{t^{\frac{1}{4}}}{\Gamma(\frac{5}{4})} \tan(\beta),$$

a result originally derived by Mullins (1963) using conventional techniques.

REFERENCES

ABEL, N. H. (1823). "Solution de quelques problèmes à l'aide d'intégrales définies." *Werke* **1**, 10.
ABEL, N. H. (1825). "Resolution d'un problème de mecanique." *Oeuvres Complètes*, Vol. 1, p. 97. Grondahl, Christiana, Norway.
ABRAMOWITZ, M., and STEGUN, I. A., eds. (1964). *Handbook of Mathematical Functions*. Nat. Bur. Stand. Appl. Math. Ser. 55. U.S. Govt. Printing Office, Washington, D. C.
ALLEGRE, J., FESTINGER, J. C., and HERPE, G. (1970). "Mesures de flux thermiques dans un ecoulement hypersonique de gaz rarefie." *Internat. J. Heat Mass Transfer* **13**, 595.
BABBITT, J. D. (1950). "On the differential equations of diffusion." *Canad. J. Res. Sect. A* **28**, 449.
BARRER, R. M. (1941). *Diffusion In and Through Solids*. Cambridge Univ. Press, London and New York.
BELAVIN, V. A., NIGMATULLIN, R. Sh., MIROSHNIKOV, A. I., and LUTSKAYA, N. K. (1964). "Fractional differentiation of oscillographic polarograms by means of an electrochemical two-terminal network." *Trudy Kazan. Aviacion. Inst.* **5**, 144.
BIORCI, G., and RIDELLA, S. (1970). "Ladder RC network with constant RC product." *IEEE Trans. Circuit Theory* **CT-17**.
BOOLE, G. (1844). "On a general method in analysis." *Philos. Trans. Roy. Soc. London* **134**, 225.
BOURLET, C. (1897). "Sur les opérations en général et les équations linéaires différentielles d'ordre infini." *Ann. Ecole Norm.* [3] **33**, 133.
CARSLAW, H. S., and JAEGER, J. C. (1947). *The Conduction of Heat in Solids*. Oxford Univ. Press (Clarendon), London and New York.
CARSON, J. R. (1926). "The Heaviside operational calculus." *Bull. Amer. Math. Soc.* **32**, 43.
CHURCHILL, R. V. (1948). *Operational Mathematics*, 2nd ed. McGraw-Hill, New York.
CIVIN, P. (1941). "Inequalities for trigonometric integrals." *Duke Math. J.* **8**, 656.
COURANT, R., and HILBERT, D. (1962). *Methods of Mathematical Physics*, Vol. II, *Partial Differential Equations*. Wiley (Interscience), New York.
CRANK, J. (1956). *The Mathematics of Diffusion*. Oxford Univ. Press, London and New York.
DAVISON, B. (1957). *Neutron Transport Theory*. Oxford Univ. Press, London and New York.
DELAHAY, P. (1954). *New Instrumental Methods in Electrochemistry*. Wiley (Interscience), New York.

DUFF, G. F. D. (1956). *Partial Differential Equations.* Univ. of Toronto Press, Toronto.
ERDÉLYI, A. (1939). "Transformation of hypergeometric integrals by means of fractional integration by parts." *Quart. J. Math. Oxford Ser.* **10**, 176.
ERDÉLYI, A. (1940). "On fractional integration and its applications to the theory of Hankel transforms." *Quart. J. Math. Oxford Ser.* **11**, 293.
ERDÉLYI, A. (1964). "An integral equation involving Legendre functions." *SIAM J. Appl. Math.* **12**, 15.
ERDÉLYI, A. (1965). "Axially symmetric potentials and fractional integration." *SIAM J. Appl. Math.* **13**, 216.
ERDÉLYI, A., MAGNUS, W., OBERHETTINGER, F., and TRICOMI, F. G. (1954). *Tables of Integral Transforms,* Vol. II. p. 181. McGraw-Hill, New York.
EULER, L. (1730). Mémoire dans le tome V des *Comment. Saint Petersberg Années,* **55**.
FADDEEVA, V. N., and TERENT'EV, N. M. (1961). *Tables of Values of the Function $w(z)$ for Complex Argument.* Pergamon Press, New York.
FELLER, W. (1952). "On a generalization of Marcel Riesz' potentials and the semi-groups generated by them." *Commun. Sem. Math. Univ. Lund Suppl.* **72**.
FRIEDMAN, B. (1969). *Lectures on Applications-Oriented Mathematics.* Holden-Day, San Francisco, California.
GAER, M. C. (1968). "Fractional Derivatives and Entire Functions." Ph.D. Dissertation, Univ. of Illinois, Urbana, Illinois.
GAER, M. C., and RUBEL, L. A. (1971). "The fractional derivative via entire functions." *J. Math. Anal. Appl.* **34**, 289.
GEMANT, A. (1936). "A method of analyzing experimental results obtained from elastoviscous bodies." *Physics (New York)* **7**, 311.
GRADSHTEYN, I. S., and RYZHIK, I. M. (1965). *Table of Integrals, Series and Products.* Academic Press, New York.
GRAHAM, A., SCOTT BLAIR, G. W., and WITHERS, R. F. J. (1961). "A methodological problem in rheology." *British J. Philos. Sci.* **11**, 265.
GRENNESS, M., and OLDHAM, K. B. (1972). "Semiintegral electroanalysis: theory and verification." *Anal. Chem.* **44**, 1121.
GRÜNWALD, A. K. (1867). "Uber 'begrenzte' Derivationen und deren Anwendung." *Z. Angew. Math. Phys.* **12**, 441.
HARDY, G. H. (1917). "On some properties of integrals of fractional order." *Messenger Math.* **47**, 145.
HARDY, G. H., and LITTLEWOOD, J. E. (1925). "Some properties of fractional integrals." *Proc. London Math. Soc.* [2], **24**, 37.
HARDY, G. H., and LITTLEWOOD, J. E. (1928). "Some properties of fractional integrals, I." *Math. Z.* **27**, 565.
HARDY, G. H., and LITTLEWOOD, J. E. (1931). "Some properties of fractional integrals, II." *Math. Z.* **34**, 403.
HEAVISIDE, O. (1892). *Electrical Papers.* Macmillan, London.
HEAVISIDE, O. (1893). "On operators in physical mathematics." *Proc. Roy. Soc.* **52**, 504; **54**, 105.
HEAVISIDE, O. (1920). *Electromagnetic Theory,* Vol. II. Benn, London. (Reprinted by Dover, New York, 1950.)
HIGGINS, T. P. (1967). *The Use of Fractional Integral Operators for Solving Nonhomogeneous Differential Equations,* Document DI-82-0677. Boeing Sci. Res. Lab., Seattle, Washington.
HILLE, E. (1939). "Notes on linear transformations, II: Analyticity of semi-groups." *Ann. of Math.* **40**, 1.

HILLE, E. (1948). *Functional Analysis and Semi-Groups*, Amer. Math. Soc. Colloq. Publ., Vol. 31. Amer. Math. Soc., New York.
HOLMGREN, H. J. (1864). "Om differentialkalkylen med indices of hvad nature sam helst." *Kgl. Sv. Vetenskapsakad. Handl.* **11**, 1.
HOLUB, K., and NEMEČ, L. (1966). "The analog method for solution of problems involving diffusion to the electrode I. Diffusion to a sphere with Langmuiran adsorption solved by analog method." *J. Electroanal. Chem. Interfacial Electrochem.* **11**, 1.
ICHISE, M., NAGAYANAGI, Y., and KOJIMA, T. (1971). "An analog simulation of noninteger order transfer functions for analysis of electrode processes." *J. Electroanal. Chem. Interfacial Electrochem.* **33**, 253.
JOHNSON, W. C. (1950). *Transmission Lines and Networks*. McGraw-Hill, New York.
JOST, W. (1952). *Diffusion*. Academic Press, New York.
KNOPP, K. (1945). *Theory of Functions*, Vol. I. Dover, New York.
KOBER, H. (1940). "On fractional integrals and derivatives." *Quart. J. Math. Oxford Ser.* **11**, 193.
KOUTECKY, J. (1953). "Theorie langsamer Elektrodenreaktionen in der Polarographie und polarographisches Verhalten eines Systems, bei welchem der Depolarisator durch eine schnelle chemische Reaktion aus einem elektroinaktiven Stoff entsteht." *Collect. Czech. Chem. Commun.* **18**, 597.
KRUG, A. (1890). "Theorie der Derivationen." *Akad. Wiss. Wien Denkenschriften Math. Naturwiss. Kl.* **57**, 151.
KUTTNER, B. (1953). "Some theorems on fractional derivatives." *Proc. London Math. Soc.* [3] **3**, 480.
LAGRANGE, J. L. (1772). "Sur une nouvelle espèce de calcul relatif à la differentiation et à l'integration des quantités variables." *Nouv. Mem. Acad. Roy. Sci. Belles-Lett. Berlin* **3**, 185; also *Ouvres de Lagrange*, Vol. 3, p. 441. Gauthier-Villars, Paris, 1849.
LEIBNIZ, G. W. (1859). *4e lettre à Wallis (1725)*. *Oeuvres* **3**.
LIOUVILLE, J. (1832a). "Mémoire: Sur le calcul des différentielles à indices quelconques." *J. Ecole Polytechn.* **13**, 71.
LIOUVILLE, J. (1832b). "Memoire sur l'integration de l'equation $(mx^2 + nx + p) d^2y/dx^2 + (qx + r) dy/dx + sy = 0$ à l'aide des differentielles à indices quelconques." *J. Ecole Polytechn.* **13**, 163.
MATSUDA, H., and AYABE, Y. (1955). "Theoretical analysis of polarographic waves. I. Reduction of simple metal ions." *Bull. Chem. Soc. Jap.* **28**, 422.
MEYER, R. F. (1960). *A Heat-Flux-Meter for Use with Thin Film Surface Thermometers*. Aeronaut. Rep. LR-279. Nat. Res. Council of Canada, Ottawa.
MIKUSIŃSKI, J. G. (1959). *Operational Calculus*. Pergamon, Oxford.
MOORE, F. K. (1964). *Theory of Laminar Flows*. Princeton Univ. Press, Princeton, New Jersey.
MULLINS, W. W. (1957). "Theory of thermal grooving." *J. Appl. Phys.* **28**, 333.
MULLINS, W. W. (1963). "Solid surface morphologies governed by capillarity." *Metal Surfaces*, p. 17. American Society of Metals, Metals Park, Ohio.
MURPHY, G. M. (1960). *Ordinary Differential Equations and Their Solutions*. Van Nostrand-Reinhold, Princeton, New Jersey.
MUSKAT, M. (1937). *The Flow of Homogeneous Fluids through Porous Media*. McGraw-Hill, New York.
NEKRASSOV, P. A. (1888). "General differentiation." *Mat. Sb.* **14**, 1.
OLDHAM, K. B. (1969a). "A new approach to the solution of electrochemical problems involving diffusion." *Anal. Chem.* **41**, 1904.
OLDHAM, K. B. (1969b). "A unified treatment of electrolysis at an expanding mercury electrode." *Anal. Chem.* **41**, 936.

OLDHAM, K. B. (1970). "On the evaluation of certain sums involving the natural numbers raised to an arbitrary power." *SIAM J. Math. Anal.* **1**, 538.
OLDHAM, K. B. (1972). "Signal-independent electroanalytical method." *Anal. Chem.* **44**, 196.
OLDHAM, K. B. (1973a). "Semiintegral electroanalysis: analog implementation." *Anal. Chem.* **45**, 39.
OLDHAM, K. B. (1973b). "Diffusive transport to planar, cylindrical, and spherical electrodes." *J. Electroanal. Chem. Interfacial Electrochem.* **41**, 351.
OLDHAM, K. B., and SPANIER, J. (1970). "The replacement of Fick's laws by a formulation involving semidifferentiation." *J. Electroanal. Chem. Interfacial Electrochem.* **26**, 331.
OLDHAM, K. B., and SPANIER, J. (1972). "A general solution of the diffusion equation for semiinfinite geometries." *J. Math. Anal. Appl.* **39**, 655.
OSLER, T. J. (1970a). "Leibniz rule for fractional derivatives and an application to infinite series." *SIAM J. Appl. Math.* **18**, 658.
OSLER, T. J. (1970b). "The fractional derivative of a composite function." *SIAM J. Math. Anal.* **1**, 288.
OSLER, T. J. (1971). "Fractional derivatives and Leibniz rule." *Amer. Math. Monthly* **78**, 645.
OSLER, T. J. (1972a). "An integral analogue of Taylor's series and its use in computing Fourier transforms." *Math. Comp.* **26**, 449.
OSLER, T. J. (1972b). "A further extension of Leibniz rule to fractional derivatives and its relation to Parseval's formula." *SIAM J. Math. Anal.* **3**, 1.
OSLER, T. J. (1972c). "The integral analog of the Leibniz rule." *Math. Comp.* **26**, 903.
POST, E. L. (1930). "Generalized differentiation." *Trans. Amer. Math. Soc.* **32**, 723.
RIEMANN, B. (1953). "Versuch einer allgemeinen Auffasung der Integration und Differentiation." *The Collected Works of Bernhard Riemann* (H. Weber, ed.), 2nd ed. Dover, New York.
RIESZ, M. (1949). "L'intégral de Riemann–Liouville et le problème de Cauchy." *Acta Math.* **81**, 1.
RITT, J. F. (1917). "On a general class of linear homogeneous differential equations of infinite order with constant coefficients." *Trans. Amer. Math. Soc.* **18**, 27.
ROBERTS, G. E., and KAUFMAN, H. (1966). *Table of Laplace Transforms*. W. B. Saunders Company, Philadelphia.
ROBERTSON, W. M. (1971). North Amer. Rockwell Sci. Center, Thousand Oaks, California. Private communication.
SCOTT BLAIR, G. W. (1947). "The role of psychophysics in rheology." *J. Colloid Sci.* **2**, 21.
SCOTT BLAIR, G. W. (1950a). *Measurements of Mind and Matter*. Dennis Dobson, London.
SCOTT BLAIR, G. W. (1950b). "Some aspects of the search for invariants." *British J. Philos. Sci.* **1**, 230.
SCOTT BLAIR, G. W., and CAFFYN, J. E. (1949). "An application of the theory of quasi-properties to the treatment of anomalous stress–strain relations." *Philos. Mag.* **40**, 80.
SCOTT BLAIR, G. W., VEINOGLOU, B. C., and CAFFYN, J. E. (1947). "Limitations of the Newtonian time scale in relation to non-equilibrium rheological states and a theory of quasi-properties." *Proc. Roy. Soc. Ser. A* **187**, 69.
SHERMERGOR, T. D. (1966). "On the use of fractional differentiation operators for describing the hereditary properties of materials." *Z. Prikl. Mech. i Tekhn. Fiz.* **6**, 118.
SOMORJAI, R. L., and BISHOP, D. M. (1970). "Integral-transformation trial functions of the fractional-integral class." *Phys. Rev. A* **1**, 1013.
TITCHMARSH, E. C. (1948). *Introduction to the Theory of Fourier Integrals*, 2nd ed. Oxford Univ. Press, London and New York.

WALL, H. S. (1948). *Analytic Theory of Continued Fractions.* Van Nostrand-Reinhold, Princeton, New Jersey.
WATANABE, Y. (1931). "Notes on the generalized derivative of Riemann–Liouville and its application to Leibniz's formula I and II." *Tôhoku Math. J.* **34**, 8.
WEYL, H. (1917). "Bemerkungen zum Begriff des Differentialquotienten gebrochener Ordnung." *Vierteljschr. Naturforsch. Gesellsch. Zürich* **62**, 296.
WIDDER, D. V. (1947). *Advanced Calculus.* Prentice-Hall, Englewood Cliffs, New Jersey.
WIENER, N. (1926). "The operational calculus." *Math. Ann.* **95**, 557.

INDEX

Page numbers followed by (n) indicate that the entry appears in a footnote on that page.

A

Abel, N. H., 2, 4, 183, 184, 219
Abel's integral equation, 2, 4, 9, 13, 183–186
Abramowitz, M., 219
Airy functions, 180
   as hypergeometrics of complexity $\frac{0}{2}$, 164
Algorithms for differintegration, 136–148
Allegre, J., 133, 202, 219
Analog differintegration, 148–154
Analytic functions, 77, 80, 116
   differintegrals of, 75
Applications of fractional calculus
   to Abel's integral equation, 183–186
   to biology, 10
   to chemical physics, 2
   to classical calculus, 181–195
   to definition of function families, 192–195
   to differential equations, 2, 8, 10, 13
   to diffusion of atmospheric pollutants, 208
   to diffusion problems, 197–218
   to electrical conduction in transmission lines, 210
   to electrochemistry, 2, 14, 15, 204
   to functional equations, 10
   to geometry, 4
   to grain boundary grooving in metals, 216–218
   to integral equations, 2, 10, 12, 13, 15
   to location of candidate solutions for differential equations, 189–192
   to mechanics, 4
   to Navier–Stokes equation, 12, 14
   to prediction of peak pollutant concentrations, 209
   to problems of elasticity, 2
   to quantitative chemical analysis, 205–207
   to rheology, 2
   to solution of Bessel's equation, 186–189
   to solution of Legendre's equation, 191
   to tautochrone problem, 2, 4, 5
   to theory of heat conduction, 202–203
   to transmission line theory, 2
   to transport problems, 2
   to wave equation, 11
Associated Legendre functions, as reducible transcendentals, 161
Ayabe, Y., 160, 221

B

Babbitt, J. D., 197, 219
Barrer, R. M., 197, 219
Basis hypergeometrics, 168–172
   complementary, 168

Basic hypergeometrics (*cont.*)
   definition of, 168
   graphs of, 168
   Laplace transformation of, 171
   relationships among, 171
Bassam, M. A., 12
Belavin, V. A., 2, 13, 219
Berg, E. J., 9
Bessel functions, 13, 97–98, 124, 125, 177–178
   as hypergeometrics of complexity $\frac{0}{2}$, 164
   as reducible transcendentals, 161
   relationships among, 98
Bessel's equation, 186–189
   solution via fractional calculus, 186–189
Beta function, *see* Complete beta functions, Incomplete beta functions
Bibliography, chronological, 3–15
Binomial coefficients, 20, 28–29, 178, 180
   of moiety argument, table of, 118
Binomial functions, 162, 174
   as hypergeometrics of complexity $\frac{1}{1}$, 99
Biology, 10
Biorci, G., 154, 219
Bishop, D. M., 2, 14, 222
Blumenthal, L. M., 10
Boole, G., x, 2, 6, 219
Boundary geometries for diffusion problems, 198
   diagrams of, 199
Bourlet, C., x, 2, 219
Brenke, W. C., 9
Buschman, R. G., 12, 13
Butzer, P., 14

C

Caffyn, J. E., 2, 222
Carslaw, H. S., 197, 219
Carson, J. R., 2, 219
Cauchy's integral formula, 1, 7, 8, 14, 54
Cayley, A., 8
Center, W., 6, 7, 9
Chain rule
   for differintegrals, 80–81
   for multiple derivatives, 36–37
Chronological bibliography on fractional calculus, 3–15

Churchill, R. V., 134, 136, 150, 219
Circuit for performing semiintegration, *see* Semiintegrating circuits
Civin, P., 53, 219
Classical calculus, 25–44
Classical derivatives, 28
Classical integrals, 29
Cole, K. S., 10
Complementary functions, 4, 5, 8
Complete beta functions, 21, 65
Complex error functions, 209
Composition rule, 48, 82–87
   for differentiable functions, 85
      tabular summary, 86
   for differintegrable units, 82–84
      tabular summary, 84
   examples illustrating failure, 83, 86–87
   for general differintegrable series, 84–85
      tabular summary, 85
   for mixed integer orders, 30–33
      failure, 32
   for noninteger orders, 63
   role in locating solutions of differential equations, 190
   use in inverting extraordinary differential equations, 155
   utility in finding differintegrals, 96
Continued fractions, 151–152
Convolution theorem for Laplace transformation, 134
Cosine integrals as reducible transcendentals, 161
Cosines, *see* Sines and cosines
Coulomb's law, 211
Courant, R., 12, 27, 50, 219
Crank, J., 197, 219
Curvature correction for current semiintegral, graph of, 206
Cyclodifferential functions, 110–112, 169(n)
   definition of, 110
Cycloid as solution to tautochrone problem, 185
   diagram of, 185

D

Davis, H. T., 8, 9, 10, 11
Davison, B., 197, 219
Dawson's integral, 163, 175

INDEX

Definite integrals, evaluation through fractional calculus, 181–183
Delahay, P., 160, 219
Delta function, see Dirac delta function
DeMorgan, A., 5
Derivatives, see Classical derivatives
Difference quotients, 28, 48, 56
 for classical derivatives, 28
Differential equations, 2, 8, 10, 13
 conversion to lower order, 190
 extraordinary, see Extraordinary differential equations
Differentiation
 to fractional order, see Differintegration
 of hypergeometrics, 40–44
 of powers, 39–40, 192
  diagram showing sign of resultant derivative, 39
 term by term, 38
Differintegrable functions, 46–47
 definition of, 47
Differintegrable and nondifferintegrable functions, figure showing examples of each, 47
Differintegrable series, 69, 70, 82, 93, 116
 definition of, 46
Differintegrable series units, 82
 definition of, 46
Differintegral operators
 analytic continuation for, 49
 comparison of definitions, 55–57
 connection with Fourier analysis, 56
 connection with Laplace transformation, 11
 defined, x, 16, 27(n)
 definition of, 45–60
  based on Cauchy's integral formula, 54, 56
   diagram of contour used, 55
  as integral transform, see Differintegral operators, Riemann–Liouville definition
  as sum, see Differintegral operators, Grünwald definition
  in terms of difference quotients, 1
  in terms of series, 1, 5, 7, 53
 differentiation of, 48, 50
 equivalence of definitions, 51–52
 general properties, 10, 11, 69–92
  behavior far from lower limit, 91

 behavior near lower limit, 90
 chain rule, 14, 80–81
 commutativity, 87
 composition rule, 82–87
 dependence on lower limit, 87–89
 differintegration term by term, 69–75
 homogeneity, 75
 inversion, 86
 Leibniz's rule, 76–79
 linearity, 69
 scale change, 75–76
 translation, 89–90
 Grünwald definition, 48, 55
 identity of definitions, 51–52
 of imaginary order, 14
 of integer order, 25–44
 modified Grünwald definition, 57
 representation for analytic functions, 57–59
 with respect to arbitrary function, 55
 Riemann–Liouville definition, 1, 6, 9, 49, 56, see also Riemann–Liouville integral
  use in evaluating integrals, 181
 some general definitions, 52–57
 summary of definitions, 59–60
 symbolism for, 45
Differintegration, 61–68, 93–114
 of Bessel functions, 97–98
 of binomial functions $[C - cx]^p$, 93–94
 of constants, 63
 of cosine function $\cos(x)$, 112
 of cyclodifferential functions, 110–112
 of Dirac delta function, 106
 of exponential functions $\exp(C - cx)$, 94–95
 of function $x^{q-1} \exp(-1/x)$, 112–113
 of function $x - a$, 63–65
 of functions $x^q/[1 - x]$ and $x^p/[1 - x]$ and $[1 - x]^{q-1}$, 95–96
 of Heaviside function, 105
  diagram illustrating, 106
 of hyperbolic function $\sinh(\sqrt{x})$, 96–97
 of hypergeometric functions, 99–102
  examples of $\frac{1}{3}$ complexity, 112
 of logarithms, 102–104
 to order one-quarter, 217
 of periodic functions, 108–110
 of piecewise-defined functions, 107

Differintegration (*cont.*)
  of powers, 63–68
    breakdown for $p \leq -1$, 67
    figure showing contour used in, 66
    signs of resulting coefficients, 68
  of sawtooth function, 107–108
    diagram of, 108
  of sine function $\sin(x)$, 112
  of sine function $\sin(\sqrt{x})$, 96–97
  of unit function, 61–63
    graphs of differintegrals, 62
  of zero function, 63
Diffusion
  on curved surface, 216–218
  in finite media, 209–215
  in planar geometry, 201–203
  in presence of sources and sinks, 207–209
  in spherical geometry, 204–207
Diffusion equation
  general form, 198
  in presence of sources and sinks, 207
  for semiinfinite geometries, 199
Dirac delta function, 105
  differintegrals of, 106
Duff, G. F. D., 27, 49, 220

E

Eigenfunction
  of differintegral operator, 122(n)
  of semidifferential operator, 122(n)
    graph of, 123
Elasticity, 2
Electrochemistry, 2, 14, 15, 204
Elliptic integrals
  of first kind, 122
    as hypergeometrics of complexity $\frac{2}{2}$, 165
    as reducible transcendentals, 161
  of second kind, 122
Erdélyi, A., 2, 11, 12, 13, 54, 55, 76, 115, 220
Error function complement, 194
Error function complement integrals, 175, 194–195
Error functions, 175
  as reducible transcendentals, 161
Errors in analog semidifferentiation, 215–216
  graph of, for various terminations, 215

Euler, L., 1, 3, 220
Euler's constant, 24
Euler's integral of second kind, 21
Exponential functions, differintegration of, 94–95
Exponential integrals, 166
  as reducible transcendentals, 161
Exponential-like functions, as hypergeometrics of complexity $\frac{0}{1}$, 162
Extraordinary differential equations, 154–157
  definition of, 154
  for describing groove depth in metals, 217
  inversion of, 155
  series solutions of, 159–160

F

Faà di Bruno's formula for differentiating composite function, 37, 80
Fabian, W., 11
Faddeeva, V. N., 209, 220
Faradaic current, semiintegral of, 205–207
  graph versus time, 206
  diagram illustrating curvature correction, 206
Faraday's constant, 204
Faraday's electrochemical law, 204
Feller W., x, 220
Festinger, J. C., 219
Fick's first law, 204
Fourier, J. B. J., 4, 5
Fourier analysis, 14
Fractional calculus
  applications of, *see* Applications of fractional calculus
  historical survey, 1–15
  symbolic methods in, 2
Fractional difference operators, 12
Fractional differential operators, *see* Differintegral operators
Fractional differentiation, *see* Differintegration
Fractional integral operators, *see* Differintegral operators
Fractional integration, *see* Differintegration

INDEX   229

for functions of more than one variable, 2
by parts, 11
Fresnel integrals, 125, 180
  as reducible transcendentals, 161
Friedman, B., x, 220
Function families, 192–195
Function synthesis, 169
  of $K = L-2$ transcendentals, 177–180
  of $K = L-1$ transcendentals, 175–176
  of $K = L$ transcendentals, 172–175
Function synthesis diagrams, see Synthesis diagrams
Functional equations, 10
Fundamental theorem of calculus, 30

G

Gaer, M. C., 13, 53, 220
Gamma function, 14, 16–24, see also Incomplete gamma functions
  asymptotic expansion, 19
  duplication formula, 18
  Gauss multiplication formula, 18
  general properties, 16–24
  for integer and half-integer arguments, 17–18
    table of values, 18
  ratios, 17, 19–20
    polynomials expressible as, 19
    graphs of, 19
  reciprocal, 17
    asymptotic representation of, 17
    graph of, 17
  recurrence formula, 16
  reflection formula, 18
  relation to beta functions, 21
  relation to binomial coefficients, 20
  relation to psi function, 23
Gauss functions, 41, 122, 165, 174
  as reducible transcendentals, 161
Gel'fand, I. M., 13
Gemant, A., 2, 67, 220
Generalized Abel equation, 186
Generalized differentiation, see Differintegration
  for operators, 2
Generalized error function complement integrals, 195
Generalized hypergeometric functions, see Hypergeometrics

Generalized integration, see Differintegration
Generalized logarithms, 163(n), 175, 192
  definition of, 193
  representation of, 193–194
Geometric factor to characterize boundary geometries in diffusion problems, 198
Gradshteyn, I. S., 220
Graham, A., 2, 220
Greatheed, S. S., 5
Greer, H. R., 7
Gregory, D. F., 5
Grenness, M., 2, 14, 144, 204, 220
Grünwald, A. K., x, 1, 7, 48, 220

H

Hadamard, J., 8
Hagstrom, K. G., 11
Hardy, G. H., 2, 8, 9, 10, 11, 220
Hargreave, C. J., 6
Heat conduction in semiinfinite planar medium, 202
Heat equation, 5
Heat flux, relation to surface temperature through semidifferentiation, 202
Heaviside, O., x, xii, 2, 8, 9, 53, 62, 220
Heaviside function, 61, 105
  graph of differintegrals, 106
Heaviside's operational calculus, see Operational calculus
Heaviside's unit function, see Heaviside function
Herpe, G., 219
Higgins, T. P., 2, 13, 220
Hilbert, D., 27, 50, 219
Hilbert transforms, 15
Hille, E., x, 220, 221
Hirschmann, I. I., 12
Holmgren, H., 1, 7, 12, 221
Holmgren–Riesz transforms, 12
Holub, K., 133, 221
Hyperbolic cosines as reducible transcendentals, 161
Hyperbolic sine integrals, 128
  as hypergeometrics of complexity $\frac{1}{3}$, 165

Hyperbolic sines
   as hypergeometrics of complexity $\frac{0}{2}$, 100
   as reducible transcendentals, 161
Hypergeometric functions, 11, 13, 14, 15
   of complexity $\frac{K}{L}$, 15
   Laplace transformation of, 171
   notation for, 15
   product with power of argument, *see* Hypergeometrics
   as reducible transcendentals, 161
Hypergeometrics, 40–41, 99–102, 129–130
   of argument $x^{1/n}$, 43, 100–101
   cancellation property of, 42
   of complexity $\frac{0}{0}$, 162
   of complexity $\frac{0}{1}$, 162
   of complexity $\frac{1}{1}$, 163
   of complexity $\frac{1}{2}$, 164
   of complexity $\frac{1}{3}$, 165
   of complexity $\frac{2}{2}$, 164–165
   convergence properties of, 165
   definition of, 162
   differentiation and integration of, 40–44
   as differintegrable series, 99
   differintegrals of, 99—102
   examples of, 162–165
   with $K > L$, 165–166
      utility for asymptotic representations, 166
   Laplace transformation, 171–172
   recurrence relations for, 42
   reduction of complexity, 166–167
   reduction to complexity $\frac{0}{L-M}$, 167
   regeneration from basis hypergeometrics, 169
   as sum of hypergeometrics, 44
      use in finding differintegrals, 101
   symbolism for, 40

I

I'a Bromwich, T. J., 9
Ichise, M , 133, 154, 221
Incomplete beta functions, 21, 41, 94, 122, 174
   as reducible transcendentals, 161

Incomplete gamma functions, 21, 41, 94, 95, 109, 158, 175
   recursion formula for, 22
   as reducible transcendentals, 161
Infinite series
   differentiation and integration of, 38
   evaluation through fractional calculus, 183
Infinite transmission line, approximation by finite transmission line, 212
Integral equations, 2, 10, 12, 13, 15
Integrals, *See* Classical integrals
Integrating circuit, 149
Integration
   to fractional order, *see* Differintegration
   of hypergeometrics, 40–44
   of powers, 39–40, 192–193
      diagram showing sign of resultant integral, 39
   term by term, 38
Intergranular groove, 216–218
   cross sectional diagram, 217
Inverse hyperbolic functions, 163
   as reducible transcendentals, 161
Inverse hyperbolic sines as hypergeometrics of complexity $\frac{2}{2}$, 165
Inverse trigonometric functions, 163
   as reducible transcendentals, 161
Iterated integrals, 37–38
   Cauchy's formula for, 38

J

Jaeger, J. C., 197, 219
Johnson, W. C., 197, 210, 221
Jost, W., 197, 221
Juberg, R. K., 15

K

Kalisch, G. K., 13
Kaufman, H., 115, 222
Kelland, P., 4, 5, 6, 7
Kesarwini, R. N., 13
Knopp, K., 49, 221
Kober, H., 2, 11, 14, 221
Kojima, T., 221
Koutecky, J., 160, 221

Koutecky function of polarography, 160
Krug, A., 1, 8, 53, 221
Kummer functions, 41, 175
 as hypergeometrics of complexity $\frac{1}{2}$, 164
 as reducible transcendentals, 161
Kuttner, B., 2, 12, 221

L

Lacroix, S. F., 4
Lagrange, J. L., 1, 3, 221
Laplace, P. S., 3
Laplace transforms, 11, 115
 of derivatives and integrals, 133
 of differintegrals, 133–136
  formula for, 134
 for performing circuit analysis, 150
 role played in solving diffusion problems, 201
Laurent, H., 8
Law of exponents, *see* Composition rule
 for operators of integral order, 3
Lebesgue class, 2, 10
Legendre functions, 122
 as hypergeometrics of complexity $\frac{2}{2}$, 164
 as reducible transcendentals, 161
Legendre's equation, 191
Leibniz, G. W., x, 1, 3, 221
Leibniz's rule, 6, 8, 10, 14, 15, 53, 62
 for derivatives, 36
 for differintegral operators
  convergence difficulties, 78–79
  integral form, 79, 182
  symmetric form, 79
  when factors are analytic, 77
  when one factor is polynomial, 77–78
 for multiple integrals, 34–35
Letnikov, A. V., 7, 8
Levy, P., 9
L'Hospital, G. A., 1, 3
Lions, J. L., 12, 14
Liouville, J., xi, 1, 4, 5, 6, 7, 8, 49, 53, 95, 191, 221
Lipschitz classes, 2
Littlewood, J. E., 2, 10, 11, 220
Liverman, T. P. G., 13
Logarithms, 163

differintegrals of, 102–104
 graphs of, 104
 generalization of, 192–194
 location in function synthesis diagram, 194
 as hypergeometrics of complexity $\frac{1}{1}$, 102
 as reducible transcendentals, 161
Love, E. R., 11, 14
Lower limit, 87–89
Lutskaya, N. K., 13, 219

M

Magnus, W., 220
Matsuda, H., 160, 221
Mechanics, 4
Mellin transforms, 11
Meyer, R. F., 133, 203, 221
Mikusínski, J. G., x, 221
Miroshnikov, A. I., 13, 219
Modified Bessel functions, 97–98, 125
 as hypergeometrics of complexity $\frac{0}{2}$, 100
Modified Struve functions, 125
 notation for, 15
Moore, F. K., 197, 221
Moritz, R. E., 8
Mullins, W. W., 217, 218, 221
Multiple integrals
 dependence on lower limit, 33–34
 symbolism for, 26, 27
Murphy, G. M., 156, 158, 159, 202, 221
Muskat, M., 197, 221

N

Nagayanagi, Y., 221
Naraniengar, M. T., 9
Navier–Stokes equation, 12, 14
Nekrassov, P. A., 8, 54, 221
Nemeč, L., 133, 221
Nerve impulse propagation, 10, 210(n)
Nessel, R., 14
Nigmatullin, R. Sh., 13, 219
Numerical differintegration
 algorithms for, 136–148

Numerical differintegration (*cont.*)
  algorithms for (*cont.*)
    based on Grünwald definition, 136
    based on linear interpolation, 140
    based on modified Grünwald definition, 137–138
    based on Riemann–Liouville definition, 138–139
    relative error in, 141–144
    weighting factors in, 144
      table of, 146–147
  approximations used in, diagram illustrating, 139
  coincidence of algorithms for, 144
  comparison of rival algorithms, 144–145
    in differintegrating $\sqrt{x}$, 145
    in differintegrating $1 - x^{\frac{3}{2}}$, 145, 148
  errors in, 141–144
    table of, 144
  nomenclature for, 136
    diagram illustrating, 137

O

Oberhettinger, F., 220
Ohm's law, 150, 211, 214
Oldham, K. B., 2, 14, 15, 133, 143, 144, 154, 160, 200, 204, 220, 221, 222
Oltramare, G., 8
Operational calculus, 9, 10, 11, 12, 13
Ordinary derivatives, *see* Classical derivatives
Ordinary differential equations, solution via fractional calculus, 186–189
Ordinary integrals, *see* Classical integrals
O'Shaughnessy, L., 8
Osler, T. J., 2, 14, 15, 54, 55, 56, 79, 81, 182, 183, 192(n), 222

P

Parseval's integral formula, 79
Peacock, G., 4, 6, 7, 8
Pennell, W. O., 10
Periodic functions, 108–110
  as hypergeometrics of complexity $\frac{0}{2}$, 163
Peters, A. S., 12
Piecewise-defined functions, 105, 107–108

Pincherle, S., 8
Poritsky, H., 11
Post, E. L., 2, 6, 8, 9, 10, 48, 222
Prabhakar, T. R., 15
Product rule, *see* Leibniz's rule
Psi function, 23, 103
  recursion formula for, 23

Q

Quasi-semiinfinite medium, condition for, 200(n)

R

Radius of convergence of differintegrable series, 70–71
Rayleigh's formulas, generalizations of, 97
Reducible transcendentals, 161
Resistive–capacitative transmission line, *see* Transmission line
Rheology, 2, 67
Ridella, S., 154, 219
Riemann, B., xi, 1, 6, 7, 49, 53, 222
Riemann–Liouville integral, 49, 55
Riemann–Liouville transforms, 115
Riemann sums, 48, 56
  for classical integrals, 29
Riemann zeta function, 142–143
  table of values, 143
Riesz, M., 2, 11, 12, 27, 49, 222
Ritt, J. F., x, 2, 222
Roberts, G. E., 115, 222
Robertson, W. M., 217, 222
Rodrigues' formula, generalization of, 192
Ross, B., xiii, 3, 15
Rubel, L. A., 14, 53, 220
Ryzhik, I. M., 220

S

Samko, S. G., 13
Sawtooth function, 107–108
  graphs of differintegrals, 108
Scale-change property, use in finding differintegrals, 100, 109
Schuyler, E., 8
Scott Blair, G. W., 2, 53, 67, 220, 222
Semicalculus, 14

INDEX

Semiderivatives, 7, 50
Semiderivatives and semiintegrals, 115–131
  definition of, 115(n)
  graphs for cos(x), 126
  graphs for sin(x), 126
  role played in solving diffusion problems, 197
  table of definitions, 115–116
  table of general properties, 116–117
  tables of
    for Bessel and Struve functions, 127–128
    for binomials, 120–121
    for complete elliptic integrals, 131
    for constants and powers, 118–119
    for exponentials and related functions, 122–124
    for generalized hypergeometric functions, 129–130
    for Heaviside function, 131
    for logarithms, 130
    for trigonometric and hyperbolic functions, 124–125
Semidifferential equations, 10, 120, 157–159
  definition of, 157
  for diffusion in planar geometry, 201
  for diffusion in semiinfinite media, 197
  for diffusion in spherical geometry, 204
  examples of, 157, 159
  occurrence in electrochemistry, 159–160
  for open-circuited finite transmission line, 213
  for short-circuited finite transmission line, 213
  techniques for solving, 157
    Laplace transformation, 159
    transformation to ordinary differential equation, 157
  used to relate current to applied voltage signal in infinite transmission line, 212
  used to relate electrochemical current to surface concentration, 205
  used to relate ground-level pollution concentrations to pollution generation rate, 208
Semidifferentiation, 14
  used to relate surface temperature to heat flux, 203

Semidifferentiation and semiintegration, analog circuits for, 148–154
Semiintegral electroanalysis, 204
Semiintegrals, 16, 185, *see also* Semiderivatives and semiintegrals
Semiintegrating circuits, 152
  accuracy of, 149
Shermergor, T. D., 2, 222
Shilov, G. E., 13
Shinbrot, M., 12, 14
Sine integrals, 128
  as hypergeometrics of complexity $\frac{1}{3}$, 165
  as reducible transcendentals, 161
Sines and cosines, *see* Hyperbolic sines
  differintegration as examples of cyclodifferential functions, 110–112
  as periodic functions, differintegration of, 110
  semiderivatives and semiintegrals, 124–126
  of $\sqrt{x}$, as hypergeometrics, 161, 163
  differintegration of, 96
Sneddon, I. N., 12, 13
Somorjai, R. L., 2, 14, 222
Spanier, J., 2, 14, 15, 200, 204, 222
Special functions of mathematical physics
  interrelations among, 79, 81
  as $K=L-2$ transcendentals, 177
Stegun, I. A., 219
Stephens, E., 10
Stirling numbers
  of first kind, 19
    table of values, 18
  of second kind, 22, 37, 81
    notation for, 15
    table of values, 22
Struve functions, 124, 125, 177–178
  as hypergeometrics of complexity $\frac{0}{2}$, 164
  notation for, 15
  as reducible transcendentals, 161
Stuloff, N., 12
Symbolic methods, 2
Synthesis diagrams, 169
  for associated Legendre functions, 173

Synthesis diagrams (*cont.*)
  Bessel–Struve line, 178
  example illustrating principles, 170
  involving steps of $\frac{1}{3}$ and $\frac{1}{4}$, 179–180
  for $K = L - 2$ transcendentals, 177–180
  for $K = L - 1$ transcendentals, 175–176
  for $K = L$ transcendentals, 172–174
  Legendre line, 173
  logarithm line, 175

T

Tautochrone, 2, 4, 5, 183
  coordinate system for, 184
  equations for, 185
Tautochrone problem, solution as a semiintegral, 185
Taylor's series, 14, 15
Techniques in fractional calculus, 133–160
Terentev, N. M., 209, 220
Term-by-term differentiation, 38, 74–75
Term-by-term differintegration, 69–75
  of arbitrary differintegrable series, 71–74
  to negative order, 71–72
  to positive order, 74–75
Term-by-term integration, 38
Titchmarsh, E. C., 79, 222
Transcendental functions
  interrelationships among, 167
  representation as hypergeometrics, 162–165
  representations of, 161–180
Transmission line
  impedance of, 212, 213
  optimum termination of, 212
  simulation by discrete components, 216
  symbolism, diagram explaining, 211
  terminations, diagram depicting four alternatives, 214
  theory, 2

Transport problems, 2
Transport in semiinfinite medium, 198–200
Transport theory, 197
Tricomi, F. G., 220

V

Veinoglu, B. C., 222
von Wolfersdorf, L., 13

W

Wall, H. S., 153, 223
Wallis, J., 3
Wallis' infinite product for π, 3
Wastchenxo, Z., 7
Watanabe, Y., 10, 79, 223
Wave equation, 11
Weakly singular intergral equation, *see* Abel's integral equation
Welland, G. V., 13
Weyl, H., 2, 8, 10, 53, 95, 223
Weyl differintegrals, 95
Weyl integral, 53
Widder, D. V., 11, 38, 223
Wiener, N., 2, 223
Withers, R. F. J., 220

Y

Young, L. C., 11

Z

Zero, differintegration of, 63
Zeta function, 142–143
Zygmund, A., 10, 11, 14